T0182169

Building Energy Modeling with OpenStudio

Larry Brackney • Andrew Parker
Daniel Macumber • Kyle Benne

# Building Energy Modeling with OpenStudio

## A Practical Guide for Students and Professionals

 Springer

Larry Brackney
National Renewable Energy Laboratory
Golden, CO, USA

Andrew Parker
National Renewable Energy Laboratory
Golden, CO, USA

Daniel Macumber
National Renewable Energy Laboratory
Golden, CO, USA

Kyle Benne
National Renewable Energy Laboratory
Golden, CO, USA

ISBN 978-3-030-08547-6          ISBN 978-3-319-77809-9   (eBook)
https://doi.org/10.1007/978-3-319-77809-9

Printed on acid-free paper

This Springer imprint is published by the registered company Springer International Publishing AG
part of Springer Nature.
The registered company address is: Gewerbestrasse 11, 6330 Cham, Switzerland

*The authors dedicate this book to their families.*

*For Larry Brackney:*
    *Raina and Nola*

*For Kyle Benne:*
    *Ashley and Desmond*

*For Daniel Macumber:*
    *Kristin, Tori, Lizzy, and Natalie*

*For Andrew Parker:*
    *Lisa and Penny*

# Foreword

The energy, environmental, and societal challenges of the twenty-first century are here; they are crystal clear; and they are daunting. Our responses to those challenges are less clear, but one component at least is obvious—we need a better building stock, one that uses less energy, provides greater comfort and security, and houses and supports the economic activity of a rapidly growing and urbanizing population.

One of the most powerful tools in our collective belts is building energy modeling (BEM), physics-based software simulation of building energy use given a description of the physical building, its use patterns, and prevailing weather conditions. BEM is a *sine qua non* tool for designing and operating buildings to the levels of energy efficiency that our future and present require. According to the AIA 2030 Commitment report, buildings designed using BEM use 20% less energy than those designed without it. BEM is also instrumental in developing and updating the codes, standards, certificates, and financial incentive infrastructure that supports energy efficiency in all building projects, including those that don't directly use BEM.

Today, every man-made artifact of any significance—from razors to race cars, potato chips to computer chips, cardboard boxes to big box stores—is prototyped virtually before being built and tested physically. Would you get on an airplane for which only two prototypes were built and whose design and systems weren't tested under a range of conditions over millions of hours of computer simulation? I hope not. Would you shave with a razor that hasn't undergone tens of thousands of hours of computer simulation? You probably would, but in actuality you do not—the margins on razors are sufficiently small that both razors and the machines that produce them have to be optimized to a degree that only computer simulation can manage. Why should razors be modeled and buildings not? Buildings provide a greater range of more important functions, and over much longer service lifetimes. The economics of building physical prototyping are more prohibitive; an owner will not pay to build and test multiple prototypes before settling on the final version. And most buildings—at least most commercial buildings—are essentially one-offs, sufficiently different from all other buildings in terms of local context project specifics so as to make high-level lessons transferrable but not full designs.

Yet, curiously and unfortunately, most buildings are still designed without the use of BEM. The same AIA 2030 Commitment report shows that only 43% of the new commercial floor space reported in 2017 used BEM during the design process. And that number is likely inflated by the fact that the Commitment is voluntary and that efficiency focused firms are over-represented in it. Anecdotal evidence suggests the real number is closer to 20%. That will not do at all.

The mission of the US Department of Energy's (DOE) Building Technologies Office (BTO) is to envision and enable a more energy-efficient building stock. BTO has identified increasing the use of BEM in building design as a high-priority high-impact vector in achieving its larger energy-savings mission. BTO's BEM program centers on the development of a state-of-the-art open-source BEM platform, which consists of the EnergyPlus BEM engine; the OpenStudio software development kit (SDK), which facilitates application development, workflow automation, and large-scale analysis; and the star of this book, the graphical OpenStudio Application.

The OpenStudio project has been a driving force in the evolution of BTO's BEM program. OpenStudio was BTO's first truly open-source software project, a strategic direction that has influenced BTO's entire BEM portfolio. Open-source is not an altruistic emergent enterprise. Successful open-source projects are funded, centrally managed, and resemble proprietary software projects in many structural and operational ways. Source control. Code reviews. Regression testing. Bug reporting and fixing. Pre-feature documentation. Post-feature documentation. The full Monty. Open-source is different in that it is transparent—anyone can inspect the algorithms and see how they are implemented under the hood. Transparency is crucial for BEM, which has many financial and regulatory use cases. And it is different in that it gives others the freedom to take the software and build upon it without paying a fee, signing a non-disclosure agreement, or even notifying the original authors. The open-source license used by OpenStudio allows derivative works to be proprietary and commercial. For OpenStudio and BTO, open-source has been an accelerant for industry and user adoption. EnergyPlus was re-released under an open-source license in 2012. Between 2012 and 2016, EnergyPlus downloads increased from about ten thousand per version to over forty thousand and the number of third-party applications using EnergyPlus grew from three to well over a dozen. The larger software industry has learned to live with and profit from open-source projects, the BEM industry is doing the same.

OpenStudio brought other modern software paradigms to BTO's BEM program, and, in some cases, to the BEM industry as a whole. A library structure with a rich application programming interface (API) created the classic three-layer engine-middleware-application architecture found in many computer systems— think of a smartphone's hardware-Android-apps setup—that fosters innovation and adoption. Online libraries that can complement centrally curated content with "crowd-sourced" content. Scripting, which OpenStudio calls Measures, improves

BEM workflow productivity, adds a layer of transparency to common BEM tasks, and transfers BEM process expertise. Scripting acts as a building block for large-scale BEM analyses. Software packages can run on commodity cloud services and allow anyone with a credit card to set up and run said analyses. Web technologies for interface development. Textbooks as BEM teaching tools!

This textbook is visually heavy on the OpenStudio Application. As every kid learns at a young age, books are much better with pictures! But like any good textbook, this one stresses universal concepts. The pictures simply hold your hand as you learn these concepts the way learning is done best, by doing. If you are a beginning student, you will learn how to use most of the basic features of OpenStudio. You will also learn the basic components, structures, and processes of BEM; weather data; thermostat set-points; constructions; schedules; thermal zoning; simple HVAC systems; output reports; and diagnostic variables. These constructs are ubiquitous and work in essentially the same way in every BEM tool. If you are a more advanced student, you will see Measures and learn how to work with the OpenStudio Parametric Analysis Tool. You will also learn how to think about model transformations algorithmically, how to organize design variants, when to use large-scale analysis and optimization and how to constrain it so that produces meaningful results, and how to reuse not only concepts but actual work from one project to the next. You will learn about OpenStudio, but you will gain a basic understanding of BEM and its processes that you can take with you to any other tool. You will get a fish *and* learn how to fish.

I hope this book helps you find a passion for BEM and that, OpenStudio or not, you decide to pursue it more seriously and maybe even professionally. Thank you for choosing to learn about this fascinating and important field.

Technology Manager for Building Energy Modeling,          Amir Roth
Building Technologies Office,
US Department of Energy
Washington, DC, USA
February, 2018

# Preface

This book is intended to provide advanced undergraduate and graduate students with an introduction to the topic of building energy modeling, simulation, analysis, and optimization. It is assumed that the reader has a basic understanding of fluid mechanics, thermodynamics, and heat transfer and is prepared to apply those fundamentals to more complex systems. The authors believe that this book will also be of value to energy efficiency professionals who are actively engaged in improving the built environment.

A number of software tools exist to model and simulate individual buildings, and many of the concepts discussed in this text are applicable to them. That said, the authors have structured this book around an open-source platform called OpenStudio and its underlying EnergyPlus simulation engine. OpenStudio is intended to facilitate the creation of many tools that make use of energy simulation to enable decision-making by diverse stakeholders at each stage of a building's lifecycle. The platform contains a number of attributes unique to energy modeling including an object-oriented data model, extensible scripting, and an analytical framework that scales from individual buildings up to portfolios. The reader will experience these and other features of OpenStudio and EnergyPlus throughout the book using a pair of example applications built with the platform.

Instructors should find the material organized in a sequence that slowly builds in complexity, enabling students to gain fundamental knowledge while applying new skills as they progress. To that end, each chapter concludes with one or more "checkpoint" exercises, so-named because they generally result in a usable model that is built upon in a subsequent exercise. The authors consider the checkpoint exercises as integral to the book's content, and we strongly urge students to work through them in their entirety.

While we believe the exercises are organized in the most appropriate order, instructors can elect to have students work through checkpoints one and three before proceeding to checkpoints two, four, and onwards without impacting prerequisites for subsequent chapters. It is really up to the instructor's preference as to whether they wish to maintain student focus on rudiments or begin mixing in sophisticated applications that expand on those fundamentals more rapidly.

The final third of the book contains advanced material that may be more appropriate for graduate students or professionals. The authors believe the material in these chapters is well suited to prepare graduate students for a variety of research tasks,[1] however we have attempted to balance that with practical applications that are approachable and will have resonance with current (or soon-to-be) professionals. Again, we have attempted to organize material in the text to provide the instructor with the greatest flexibility in adapting it to the needs of their students.

Whether you are a student, instructor, or practicing energy efficiency professional, it is our sincere hope that this book will be useful to you in learning about how the energy performance of the built environment is modeled and analyzed. Perhaps more importantly, it is our hope that you will use what you learn to make a real difference in how buildings impact the environment that we all share.

Golden, CO, USA                                                    Larry Brackney
                                                                   Andrew Parker
                                                                 Daniel Macumber
                                                                      Kyle Benne

---

[1] The authors and their colleagues at NREL regularly use OpenStudio on a large number of research projects for the US Department of Energy and other clients.

# Acknowledgments

The authors thank their colleagues who continue to develop and improve OpenStudio and related applications:

**NREL**: Dr. Brian Ball, Willy Bernal, Dr. Jason DeGraw, Dr. Katherine Fleming, Luigi Gentile Polese, David Goldwasser, Dr. Elaine Hale, Scott Horowitz, Rob Guglielmetti, Ry Horsey, Edwin Lee, Nick Long, Noel Merket, Noah Pflaum, Ben Polly, Joe Robertson, Marjorie Schott, Alex Swindler, and Evan Weaver
**Affiliated Engineers**: Arif Hanif and James McNeil
**Ambient Energy**: Eric Ringold and Matt Steen
**ANL**: Dr. Ralph Muelheisen
**Autodesk**: Krishnan Gowri
**Devetry**: Katie Noland, Brian Schiller, and Allan Wintersieck
**Group14**: Xia Fang and David Heinicke
**IIIT**: Subhash Jegi
**Independent Consultants**: Julien Marrec and Jason Turner
**LBNL**: Yixing Chen, Dr. Tianzhen Hong, Baptiste Ravache, and Kaiyu Sun
**NORESCO**: Matt Leach
**NRCan**: Jeff Blake, Kamel Haddad, Phylroy Lopez, Maria Mottillo, and Padmassun Rajakareyar
**NRCC**: Iain Macdonald
**ORNL**: Mark Adams, Piljae Im, Mini Malholtra, and Jibo Sanyal
**Pennsylvania State University**: Chong Zhou
**PNNL**: Yan Chen, Bing Liu, YuLong Xie, and Jian Zhang
**PSD**: Chris Balbach

OpenStudio is built on the excellent foundations of the EnergyPlus and Radiance simulation engines and would not be possible without the contributions and hard work of those development teams as well.

We appreciate the support of NREL management including Bill Livingood and Dr. Chuck Kutscher.

Thanks to the following who provided constructive feedback on the book concept and various drafts:

Daniel Bishop, Ramin Faramarzi, Dr. Nelson Fumo, Dr. Gregor Henze, Dr. Sammy Houssainy, Dr. Susan Krumdieck, and our editor Denise Penrose.

And finally, we thank the US Department of Energy and Dr. Amir Roth and Dr. Dru Crawley for funding and support of both the EnergyPlus and OpenStudio projects.

# Contents

The original version of this book was revised. A correction to this book can be found at
https://doi.org/10.1007/978-3-319-77809-9_10

# Abbreviations and Acronyms

| | |
|---|---|
| AEDG | Advanced Energy Design Guide |
| AIA | American Institute of Architects |
| AMI | Amazon Machine Image |
| AMY | Actual Meteorological Year |
| ANL | Argonne National Laboratory |
| API | Application Program Interface |
| ASHRAE | American Society of Heating, Refrigeration, and Air-Conditioning Engineers |
| AWS | Amazon Web Service |
| BCL | Building Component Library |
| BEM | Building Energy Modeling |
| BLAST | Building Loads Analysis and System Thermodynamics Program |
| CAD | Computer Aided Design |
| CBECS | Commercial Building Energy Consumption Survey |
| CDD | Cooling Degree Day |
| CEC | California Energy Commission |
| CEUS | California End-Use Survey |
| CLI | Command Line Interface |
| COP | Coefficient of Performance |
| CPUC | California Public Utility Commission |
| CVRMSE | Coefficient of Variation of Root Mean Squared Error |
| CW | Cold Water |
| DCV | Demand-Controlled Ventilation |
| DOAS | Dedicated Outdoor Air System |
| DOD | Department of Defense (United States) |
| DOE | Department of Energy (United States) |
| DOE | Design of Experiments |
| DX | Direct Expansion |
| EC-2 | Elastic Cloud Computing |
| EE | Energy Efficiency |

| | |
|---|---|
| EMS | Energy Management System |
| EPW | EnergyPlus Weather (File) |
| ERV | Energy Recovery Ventilation |
| EUI | Energy Use Intensity |
| HID | High-Intensity Discharge |
| HW | Hot Water |
| IDE | Integrated Development Environment |
| IDF | Input Data File |
| gbXML | Green Building Extensible Markup Language |
| GLHEPro | Ground Loop Heat Exchanger design tool Pro |
| GSHP | Ground Source Heat Pump |
| GUI | Graphical User Interface |
| HDD | Heating Degree Day |
| HTML | Hypertext Markup Language |
| HUD | Department of Housing and Urban Development (United States) |
| HVAC | Heating, Ventilation, and Air-Conditioning |
| IES | Illuminating Engineering Society |
| JSON | JavaScript Object Notation |
| LBNL | Lawrence Berkeley National Laboratory |
| LEED™ | Leadership in Energy and Environmental Design |
| LHS | Latin Hypercube Sampling |
| LPD | Lighting Power Density |
| NBS | National Bureau of Standards (United States) |
| NBSLD | National Bureau of Standards Load Determination program |
| NIST | National Institute of Standards and Technology (United States) |
| NMBE | Net Mean Bias Error |
| NRCan | Natural Resources Canada |
| NREL | National Renewable Energy Laboratory |
| NSGA2 | Non-dominated Sorting Genetic Algorithm 2 |
| OA | Outdoor Air |
| ORNL | Oak Ridge National Laboratory |
| OS | OpenStudio |
| OSA | OpenStudio Analysis (JSON File) |
| OSM | OpenStudio Model (File) |
| OSW | OpenStudio Workflow (JSON File) |
| PAT | Parametric Analysis Tool |
| PLR | Part Load Ratio |
| PNNL | Pacific Northwest National Laboratory |
| PSO | Particle Swarm Optimization |
| PTAC | Packaged Terminal Air Conditioning |
| RECS | Residential Energy Consumption Survey |
| RGENOUD | R-GENetic Optimization Using Derivatives |
| RTU | Rooftop Unit |
| SAT | Supply Air Temperature |
| SDK | Software Development Kit |

| SPEA2 | Strength Pareto Evolutionary Algorithm 2 |
| TMY | Typical Meteorological Year |
| UID | Unique Identifier |
| URL | Uniform Resource Locator |
| VAV | Variable Air Volume |
| WWR | Window-to-Wall Ratio |
| XML | Extensible Markup Language |

# List of Figures

# List of Tables

# Chapter 1
# Introduction to Building Energy Modeling

## 1.1 Why Modeling?

There is good reason that so much attention is paid to the concept of mathematical model in engineering and physics curriculum. Simple regressions derived from empirical data, differential equations based on first-principles, or detailed computational fluid dynamic simulations each provide an analytical framework that yields insight into the behavior of physical systems. In turn, those insights can lead to design decisions that have real impact on safety, cost, and performance of the cars we drive, the power grids that deliver our electricity, and the energy efficiency of the buildings we live and work in.

The cost/benefit of modeling has varied across markets and over time. A few examples include:

- **1960s Aerospace**: The aerospace industry was early to embrace model-based design in order to manage the incredible cost of prototyping aircraft while protecting the lives of test pilots.
- **1970s Automotive**: Increasingly stringent fuel efficiency and emissions standards, coupled with reliance on complex electronic controls drove engine and car manufacturers to adopt sophisticated model-centric processes to minimize development cost and time to market.
- **1980s Financial**: While statistical analysis had long been used to assess trends and risk in financial markets, widespread integration of computers into financial transactions put greater pressure on real-time modeling and analytics to maximize profit in both short and long terms.
- **2000s Power**: Beside the issues of load growth and emissions reduction, the power sector was faced with myriad challenges ranging from increased penetration of renewable energy resources to the introduction of demand response strategies that introduced volatility to the grid. Increasingly sophisticated models of generation, transmission, distribution, and demand systems were required to plan capital expenditures, schedule power reserves, etc.

© Springer International Publishing AG, part of Springer Nature 2018
L. Brackney et al., *Building Energy Modeling with OpenStudio*,
https://doi.org/10.1007/978-3-319-77809-9_1

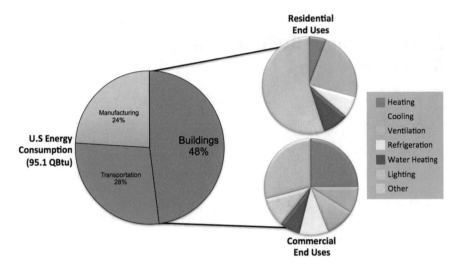

**Fig. 1.1** U.S. Energy consumption by sector with end use breakdowns. (Data source U.S. Energy Information Administration 2012)

That is not to say these (and other) sectors did not make use of mathematical models earlier – they did. This brief list is meant to point out significant historical events such as the space race, the 70s fuel crisis, advent of the personal computer, etc. that were significant drivers towards the adoption of rigorous mathematical modeling to meet market challenges. Fortuitously, these needs were enabled by improvements in the computing capability required to perform increasingly sophisticated analysis.

So, what of the topic of this book, the built environment? In a 2012 U.S. Energy Information Administration survey buildings consumed nearly half of the 95.1 Quadrillion BTUs of energy produced in the United States. Figure 1.1 shows overall consumption by sector along with a breakdown of end uses in both residential and commercial buildings.

Given the increased urgency in curbing global greenhouse gas emissions, reducing the carbon footprint of new and existing buildings has become a priority in many jurisdictions. This is evident in increasingly stringent energy efficient building codes and standards,[1,2] voluntary performance certification programs like LEED™ (Leadership in Energy and Environmental Design),[3] the Architecture 2030

---

[1] American Society of Heating, Refrigeration, and Air-Conditioning Engineers (ASHRAE) Standards and Guidelines Overview (https://www.ashrae.org/standards-research--technology/standards--guidelines).

[2] California Energy Commission (CEC) Title 24 Energy Efficiency Standards (http://www.energy.ca.gov/title24/).

[3] Leadership in Energy and Environmental Design (LEED™) Certification Overview (https://www.usgbc.org/leed).

Commitment,[4] and a host of utility incentive programs that are driving adoption of energy efficiency (EE) technologies. As in other industries, system modeling is a valuable tool for transforming the built environment. That said, why aren't analytical design tools already in widespread use?

Unlike other sectors, the buildings industry does not generally benefit from an "economy of scale." A car manufacturer may amortize investment in analytical capability across many products sold. On the other hand, buildings are usually "one offs" produced by a diverse group of stakeholders including architects, engineering firms, construction companies, owners, and occupants – none of which could would recoup investment during a project. Assuming that foundational modeling capability even existed, the perceived value of building energy modeling (BEM) varies wildly across these stakeholders, and a prospective building owner may not be inclined to procure a model for their project even though it could reduce their long-term operating expenses.

## 1.2  A Brief History of Building Energy Modeling and Simulation

The path to what we now recognize as BEM and simulation running on personal computers or high performance computing clusters has its roots in the 1960s. Digital computing had evolved to the point that ordinary and partial differential equations representing exterior and interior heat transfer involving buildings could be solved numerically. Unfortunately, even the most rudimentary analysis was restricted to academics or federal employees with access to the mainframe computers of the day. The U.S. Postal Service is recognized as one of the first federal agencies to make significant investment in dedicated software that could assess the thermal performance of buildings.[5]

Additional federal agencies began making their own investments in building energy simulation codes in the early 1970s, largely in response to geopolitical and economic turmoil of the time. The U.S. National Bureau of Standards (NBS), now known as the National Institute of Standards and Technology, created NBSLD, the National Bureau of Standards Load Determination program.[6] NBSLD was a groundbreaking FORTRAN program successfully used to study the growing building energy consumption problem for NBS, the Department of Housing and Urban Development (HUD), and the Department of Defense (DOD).

Around this time, DOD adapted NBSLD's source code into a new simulation software that would be called the Building Loads Analysis and System Thermodynamics program or BLAST.[7] In parallel, the Department of Energy (DOE)

---

[4] American Institute of Architects (AIA) 2030 Commitment Overview (https://www.aia.org/resources/6616-the-2030-commitment).

[5] Lau and Ayres (1979).

[6] Kusuda (1976).

[7] Hittle (1977).

elected to adapt the Postal Service software into its own FORTRAN code that would come to be known as DOE-2[8] developed at Lawrence Berkeley National Laboratory (LBNL). In 1998, DOD discontinued BLAST and DOE halted support for DOE-2, although DOE-2 development would continue for a number of years with funding from the California Public Utilities Commission (CPUC).

In 1997 and 1998, DOE co-authored several papers, including "Beyond BLAST and DOE-2: EnergyPlus, a New-Generation Energy Simulation Program.[9]" These described in some detail, a technical comparison of the approaches used by both BLAST and DOE-2 along with DOE's roadmap for creating new software that would draw on the best of both predecessors. The new simulation engine would be coded from scratch and known as EnergyPlus. Originally written in FORTRAN, EnergyPlus was converted to C++ in 2014 through a code-contribution from Autodesk. EnergyPlus is made available under a permissive BSD-style open source license in collaboration with multiple national laboratories, universities, contractors, and companies[10] like Trane and Carrier.

Part of the original EnergyPlus design philosophy was that it would be a simulation "engine" only, taking as input a text file and returning simulation results. EnergyPlus text Input Data Files (IDFs) for buildings of any significant complexity were tens of thousands of lines long, requiring precise specification of building geometry, constructions, individual thermal loads, detailed heating ventilation and air conditioning (HVAC) interconnections, and more. Crafting these files by hand was a time consuming and error prone undertaking. It was always DOE's intention that third-party software developers would step up to create graphical user interfaces (GUI) that would author IDFs. Unfortunately, with DOE-2 and its freely available eQUEST GUI available, along with the advent of new simulation engines like ESP-r and ApacheSIM and their own interfaces, DOE investment in a state of the art engine like EnergyPlus was not reaching the marketplace.

In 2008, a researcher at the National Renewable Energy Laboratory (NREL), Peter Ellis, created what he referred to as the OpenStudio SketchUp Plug-In.[11] Google SketchUp, as it was then known, was free and used by a significant number of architects for early concept design. It stood to reason that creating a Plug-In to convert building geometry from SketchUp into IDF would tackle one of the obstacles for modeling and analysis with EnergyPlus. That solved one part of the IDF authoring problem, but what about all of the other data required as input for EnergyPlus? If making the creation of a single model was easier, would it be possible to automate the creation of hundreds or thousands of design alternatives without the use of a laboratory or university super computer? Moreover, was there a way to help third-party software developers unleash the power of EnergyPlus more easily? The next step in the evolution of "OpenStudio" sought to answer these questions and more.

---

[8] https://buildings.lbl.gov/sites/default/files/lbnl-18046.pdf.

[9] Crawley et al. (1998).

[10] http://energyplus.net.

[11] The SketchUp Plug-In is now called "Euclid" and is available here: http://bigladdersoftware.com/projects/euclid.

## 1.3  What Is OpenStudio?

For an increasing number of people, it will be hard to recall a time when one didn't use an "app store" to find and download software for personal computers and mobile devices. The advent of operating systems, associated software development kits (SDKs), and app store-like distribution mechanisms has transformed the pace with which software innovations are brought to market. Faced with slow uptake of EnergyPlus, DOE wondered if a similar approach to software development might spur development of BEM-based applications that supported the built environment's many stakeholders.

In 2010, NREL released version 0.1 of a new version of OpenStudio that had been re-envisioned as an SDK that aims to reduce the time and expense of developing new BEM applications.[12] Subsequent quarterly releases increased coverage of EnergyPlus capability, supported DOE's Radiance daylighting analysis engine, added example applications, and introduced entirely new BEM paradigms. OpenStudio has also served as an effective collaboration platform, coordinating contributions from colleagues at Argonne (ANL), Lawrence Berkeley (LBNL), Oak Ridge (ORNL), and Pacific Northwest National Laboratories (PNNL), Pennsylvania State University, National Resources Canada (NRCan), Group 14 Engineering, and others.

Like EnergyPlus, OpenStudio is offered under a BSD-style open source license that allows companies to create and sell their own derivative works built with the SDK. It runs on Windows, Mac, and Linux, and has been used to create web and server applications as well. While the SDK itself is written in C++, code "bindings" allow it to be invoked from other languages including Ruby, C#, Python, and Javascript. As of this writing, nearly two-dozen applications have been produced using the SDK by NREL, other national laboratories, and private sector developers. A few are highlighted in Fig. 1.2, catering to a diversity of end users including building and portfolio owners, architects, engineers, policy makers, and utilities.

In addition to its role in expediting the development of innovative applications, OpenStudio has borrowed a few key concepts from other software domains, adding them to the BEM discourse. We will learn more about them throughout the book, but they are worth introducing here:

- **Application Programming Interface (API)** – Computer systems tend to be built in "layers". Each layer implements some level of functionality and then presents that functionality to the layer above using a convenient API. The API is essentially a software "contract" that relieves developers of higher-level layers from lower-level implementation details and allows layers to evolve largely independently. The OpenStudio SDK contains a rich API that is the basis for creating OpenStudio Measures or full-blown applications. The OpenStudio API also allows client applications to evolve independently from EnergyPlus by insulating the former from changes in the latter. The API is discussed in Chap. 9.

---

[12] https://www.openstudio.net.

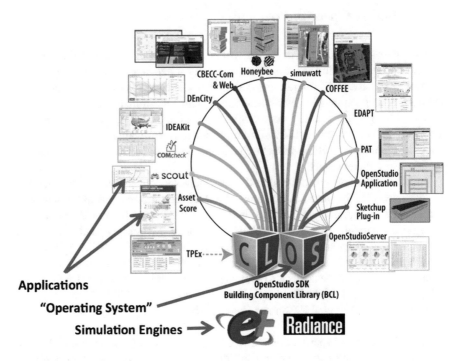

**Fig. 1.2**  OpenStudio as a BEM operating system. (*Credit: Marjorie Schott*)

- **Hierarchy and Inheritance** – The term "object-oriented" is somewhat broad and encompasses several characteristics. However, one of the main characteristics is the notion of an object hierarchy with parent-child relationships, in which more-specific child objects "inherit" attributes and capabilities from more-general parent objects. OpenStudio makes extensive use of hierarchy and inheritance to improve the efficiency of creating and modifying models. The concepts of hierarchy and inheritance as they apply to BEM will be introduced in Chap. 2.
- **Scripting and Measures** – The OpenStudio API may be invoked from the Ruby scripting language. The OpenStudio SDK itself can execute Ruby programs, essentially creating an extension and automation facility for the SDK that is similar to Visual Basic for Excel. Specially structured OpenStudio Ruby scripts are called Measures because the most common use case is applying an energy efficiency measure (EEM) to a building model to improve simulated performance in the same way the EEM is applied to a building to improve actual performance. However, as we will learn in Chap. 6, OpenStudio Measures are much more powerful and have become one of the core value propositions of the platform.
- **Cloud Computing** – The availability of commodity cloud computing has been game changing for other sectors, and it should be no different for BEM. OpenStudio

makes it easy for anyone[13] to leverage cloud computing for large-scale sampling, optimization, and other analyses. OpenStudio Measures are a key to leveraging cloud computing as they provide a systematic way of defining a large simulation space. We will explore this capability first-hand in Chap. 7.

- **Shared Content and Crowd-Sourcing** – The "open" in OpenStudio comes from open source. Modern open source software projects are largely about freedom to create and commercialize derivative works, but many also try to create a community that encourages—or at least enables—sharing of knowledge and created content. OpenStudio does this via the Building Component Library (BCL), an online repository for OpenStudio content including Measures. The BCL is first introduced in Chap. 2 but will pop up in subsequent Chapters as well.

## 1.4    Overview of Book Structure

The majority of this text is organized around one of the first example applications built using the OpenStudio SDK, the OpenStudio Application. The "App" supports construction, simulation, and review of individual building energy models. Figure 1.3 is a typical screenshot of the App annotated to highlight its workflow-centric design. Tabs along the left-hand side of the window are generally placed in the order they are used, although not all Tabs are necessarily required for modeling a building. Some Tabs are also broken down into Sub-Tabs identified along the top of the window. In general, the first (left most) Sub-Tab will be most frequently used, while subsequent sections are used in specialized circumstances.

Like the OpenStudio Application, this book is organized according to steps commonly used in a BEM workflow. The Application's Tabs and related Chapters are as follows:

**Site** – Specify weather, life cycle costs and utility bills (*Chap.* 2)
**Schedules** – Define schedules applied to building loads (*Chap.* 3)
**Constructions** – Specify materials, construction assemblies, and sets (*Chap.* 2)
**Loads** – Define individual building loads (*Chap.* 3)
**Space Types** – Create profiles for how spaces are occupied ( *Chap.* 3)
**Geometry** – Define the building exterior and interior geometries (*Chap.* 2)
**Building** – Assign building level defaults and exterior items (*Chaps.* 2, 3, *and* 8)
**Spaces** – Assign profiles to individual spaces (*Chap.* 3)
**Thermal Zones** – Group spaces into Thermal Zones and assign Zone Equipment (*Chaps.* 4 *and* 5)

---

[13] Previously, large-scale BEM analysis was the purview of laboratory or university researchers with access to high performance computers. No small engineering firms can run hundreds of simulations in the space of a few minutes for under $10.

■ **HVAC** – Specify heating, cooling, and water systems for the building (*Chaps.* 4 *and* 5)

■ **Variables** – Specify additional simulation reporting variables (*Chaps.* 2, 3, 4, *and* 5)

■ **Simulation Settings** – Customize simulation settings (*Chap.* 8 *and Appendix A*)

■ **Measures** – Assign OpenStudio Measure scripts to a workflow (*Chaps.* 6 *and* 9)

■ **Run Simulations** – Perform a single energy simulation (*Chaps.* 2, 3, 4, 5, *and* 6)

■ **Reports** – Review simulation results for a single energy simulation (*Chaps.* 2, 3, 4, 5, *and* 6)

Simulating individual buildings is certainly useful, but the real power of modeling is in the ability to perform comparative analysis of many candidate designs. A second example OpenStudio application aimed at comparative analysis of multiple models is called the Parametric Analysis Tool, or "PAT" for short (Fig. 1.4).

PAT takes one or more models developed using the OpenStudio Application and modifies them using OpenStudio Measures for small or large-scale analysis. PAT is organized according to a tabular workflow as follows:

■ **Analysis** – Specify the type of analysis to be performed (*Chaps.* 6 *and* 7)

■ **Design Alternatives** – Manually define Design Alternatives for small studies (*Chap.* 6)

■ **Outputs** – Specify key analysis outputs for large-scale analysis and visualization (*Chap.* 7)

■ **Run Simulations** – Manage local or cloud-based analysis (*Chaps.* 6 *and* 7)

■ **Compare Results** – Compare results from small studies (*Chap.* 6)

■ **Analysis Server** – Manage and interact with large-scale cloud analyses (*Chap.* 7)

Each Chapter begins with multiple sections describing one or more concepts corresponding to the App or PAT Tabs. Focus then shifts to a series of exercises that are intended to practically demonstrate those same concepts. These are referred to as "Checkpoint" exercises because they are intended to build upon each other successively. That is, Chap. 4's exercise picks up where Checkpoint Four left off at the end of Chap. 3. Each Chapter concludes with some suggestions for additional exercises that allow the reader to explore concepts in greater depth.

By following the material and performing the exercises in each Chapter, the reader will learn the rudiments of modeling as applied by many BEM tools, and become familiar with the OpenStudio SDK using example applications. Chapter 9 introduces the reader to direct application of the SDK, however it is not intended as a comprehensive guide. Readers requiring more depth on these topics are referred to the resources section of the Appendix for links to in-depth online documentation.

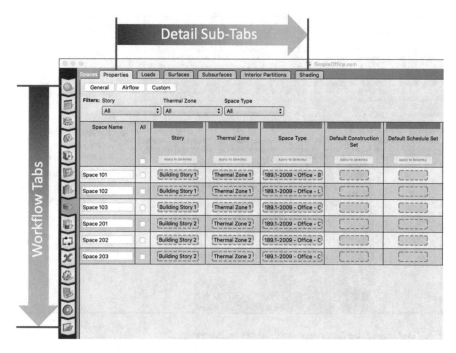

**Fig. 1.3**  OpenStudio Application workflow and detail

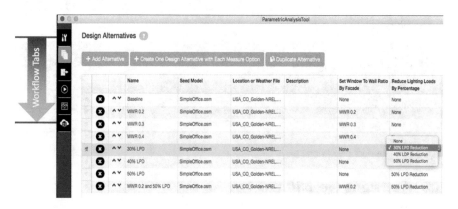

**Fig. 1.4**  OpenStudio Parametric Analysis Tool (PAT) workflow

**Fig. 1.5**  OpenStudio home page (http://openstudio.net)

**Fig. 1.6**  OpenStudio download page

## 1.5   Installing OpenStudio

The remaining Chapters assume that you have installed OpenStudio and its example Applications on your computer. OpenStudio is a free download available at http://openstudio.net (Fig. 1.5). The OpenStudio website includes downloads, documentation, tutorials, links to instructional YouTube videos, and more.

Clicking the Downloads link at the top of the webpage takes you to the Download page shown in Fig. 1.6. Links for Windows, Linux, and Mac downloads of the latest major release of OpenStudio are available on this page. Major releases are made

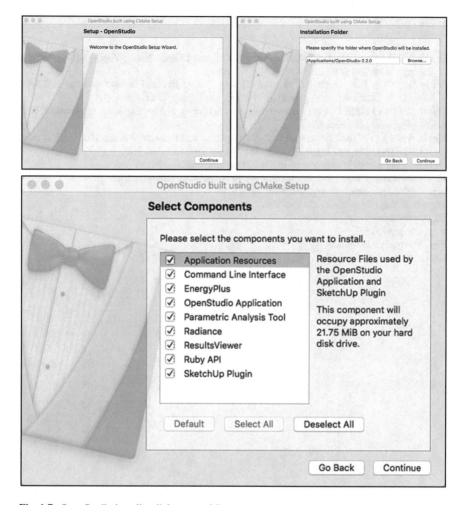

**Fig. 1.7** OpenStudio installer dialogs on a Mac

every 3 months. Minor release packages are provided more frequently on the Developers page but are not recommended for users as they may include features that are under development and unstable. Select the correct installer for your system and download the package.

Launch the downloaded package and follow the instructions to install OpenStudio (Fig. 1.7). To perform all of the activities described in this book, you will need to install all of the components offered except for the SketchUp Plugin, which is optional. Following a successful installation, it's time to begin learning about BEM and OpenStudio!

# References

Crawley D, Lawrie L et al (1998) Beyond BLAST and DOE-2: EnergyPlus, a new-generation energy simulation program. ACEEE Summer Study

Hittle D (1977) The building loads analysis and system thermodynamics program, BLAST, U.S. Army Construction Engineering Research Laboratory, Champaign, IL

Kusuda T (1976) NBSLD, the computer program for heating and cooling loads in buildings, NBS Building Science Series 69

Lau H, Ayres J 1979 Building energy analysis programs. In: Proceedings of the 11th conference on winter simulation, vol 1, pp. 283–289, San Diego, CA

https://www.aia.org/resources/6616-the-2030-commitment

https://www.ashrae.org/standards-research--technology/standards--guidelines

http://bigladdersoftware.com/projects/euclid

https://buildings.lbl.gov/sites/default/files/lbnl-18046.pdf

http://www.energy.ca.gov/title24

http://energyplus.net

https://www.openstudio.net

https://www.usgbc.org/leed

# Chapter 2
# Building Envelope Specification

## 2.1  Building Envelope

The most basic definition of a building is a man-made structure that isolates the interior from the outdoor environment. The portions of the building that separate the building's interior from the outdoor environment (e.g. walls, roofs, floors) are often referred to as the building envelope. The envelope protects the interior from rain, snow, wind, and excessive heat or cold; helping to make the interior a safe, comfortable, and productive environment for its occupants. Often, a building's interior is conditioned with Heating, Ventilation and Air Conditioning (HVAC) to maximize occupant comfort. There are many important considerations when designing a building envelope. The envelope must be sufficiently strong to support itself. It must effectively keep water or other unwanted environmental materials from damaging the building or its contents. It must be secure enough to keep unwanted pests (or people) out of it. It must be visually appealing. These aspects are all very important and there are numerous texts devoted to each of them. As this book is devoted to building energy modeling our focus will be on the transfer of energy through the building envelope.

## 2.2  Weather

As noted weather personality Willard Scott once said, "Everyone complains about the weather, but nobody ever seems to do anything about it." It should come as no surprise that weather drives a significant portion of energy transfer into and out of

The original version of this chapter was revised. A correction to this chapter can be found at https://doi.org/10.1007/978-3-319-77809-9_10

**Electronic Supplementary Material:**  The online version of this chapter (https://doi.org/10.1007/978-3-319-77809-9_2) contains supplementary material, which is available to authorized users.

L. Brackney et al., *Building Energy Modeling with OpenStudio*,
https://doi.org/10.1007/978-3-319-77809-9_2

a building. An office building in Alaska will be subject to very different environmental conditions than an office building in Florida. It is also well known that, while the actual weather conditions occurring at any given time are difficult to predict, a location's general climate may be described in a meaningful way.

Because weather varies from year to year, a methodology for combining measured weather data from multiple years into what is referred to as a "Typical" Meteorological Year (TMY) has been developed.[1] TMY data attempts to represent both the annual average weather as well as a range of weather extremes that a given location experiences. This makes TMY data more useful in predicting future energy use than Actual Meteorological Year (AMY) data for a particular year. TMY data for many locations may be downloaded in EnergyPlus Weather (EPW) format from https://energyplus. net/weather. EPW files are a key input for any OpenStudio Model, representing the ambient conditions a building is exposed to.

The American Society of Heating, Refrigerating, and Air-Conditioning Engineers (ASHRAE) categorizes a location's climate into one of several climate zones based on TMY data for that location.[2] ASHRAE climate zones are codified with a climate zone number ranging from 0 for extremely hot through 8 for sub-arctic along with a sub-type letter: **A** for Moist, **B** for Dry, and **C** for Marine. Each location's climate zone number is a function of Heating Degree Days (HDD) and Cooling Degree Days (CDD) calculated from TMY data for that location.

HDD are calculated by summing the difference between a base temperature (typically 65 °F) and the average hourly outdoor air temperature over an entire year, discarding any hours for which the outdoor air temperature is greater than the base temperature. Assuming that a typical building requires no mechanical heating when outdoor air temperatures are above the base temperature this gives a rough metric related to how much heating energy will be required. CDD are calculated in a similar manner with a different base temperature for cooling (typically 50 °F).

$$HDD = \sum_{i=1}^{8760} \frac{MAX\left(T_{base\ heating} - T_i, 0\right)}{24}$$

$$CDD = \sum_{i=1}^{8760} \frac{MAX\left(T_i - T_{base\ cooling}, 0\right)}{24}$$

The climate zone subtype is related to the location's humidity and is a function of the average annual rainfall as well as the outdoor air temperatures. Using these definitions, the climate zone designation for all locations in the US is illustrated in Fig. 2.1. A representative city has been chosen for each climate zone, and it is often assumed that TMY data for these cities are representative of the entire climate zone when performing large-scale analyses involving prototypical buildings.

A location's climate zone designation is useful for understanding the temperatures that a building experiences. However, there are many more aspects to a loca-

---

[1] Wilcox and Marion (2008).
[2] ANSI/ASHRAE (2013).

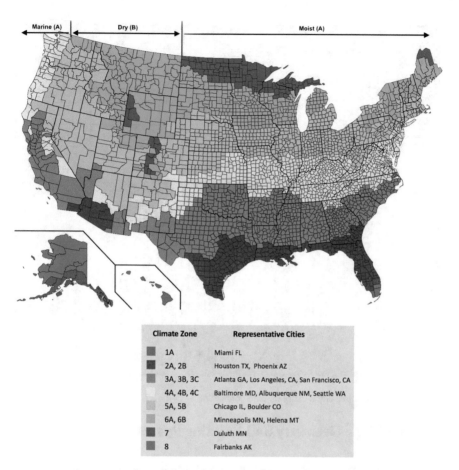

**Fig. 2.1** Climate zone locations in U.S

tion's environment including altitude, wind speed and direction, psychrometric conditions, and solar radiation incident on the surface of the Earth (solar insolation) that are often captured in weather data files. Tools like DView (included with OpenStudio), Climate Consultant from UCLA (Fig. 2.2), Elements from Big Ladder Software, or Ladybug for Grasshopper are useful for exploring TMY data for a particular location in detail.

Additional weather information useful for energy modeling is also found in Design Day (DDY) files, which are freely available for many locations at https:// energyplus.net/weather. Data in DDY files describe extreme climate conditions expected for a particular location. DDY files are frequently used when sizing HVAC systems since they must be capable of keeping the building comfortable during extreme heating, cooling, humidification, and dehumidification conditions. We shall revisit the topic of design days and HVAC sizing in Chap. 4.

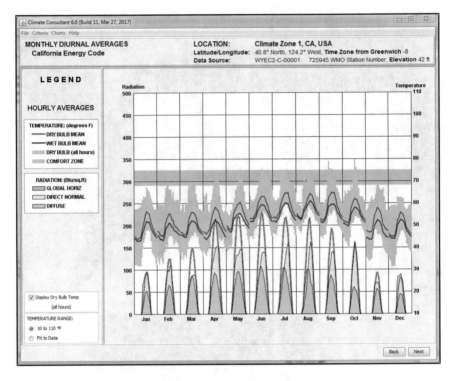

**Fig. 2.2** TMY data inspected using climate consultant

## 2.3   Envelope Geometry and Building Spaces

There are number of ways to develop envelope and interior geometry for an OpenStudio Model. The floor plan editor integrated within the OpenStudio Application may be used to develop a two-dimensional floor plan for each building story (Fig. 2.3). The OpenStudio Plug-In for Trimble SketchUp can also be used to modify detailed building geometry in three dimensions (Fig. 2.4). Third party Computer Aided Drafting (CAD) tools capable of exporting geometry in Green Building Extensible Markup Language (gbXML) format may also be used since OpenStudio can import gbXML files. Finally, the OpenStudio API, discussed in Chap. 9, may be used to procedurally create geometry for an OpenStudio Model. While this book focuses primarily on the free and open source floor plan editor integrated with the OpenStudio Application, the general concepts discussed in this Chapter apply to the geometry of any OpenStudio Model, regardless of the source.

The interior geometry of an OpenStudio Model is composed of distinct (non-overlapping), three-dimensional volumes called Spaces. It is important to note that Spaces are an OpenStudio abstraction not presently shared with EnergyPlus. OpenStudio Spaces are useful for specifying programmatic activities within a building (e.g. Office, Classroom, etc.) and their corresponding energy loads. A shared

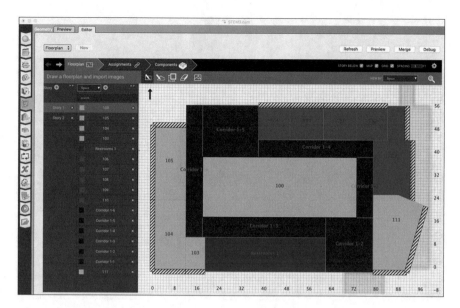

**Fig. 2.3** Building envelope drawn with the OpenStudio Application floor plan editor

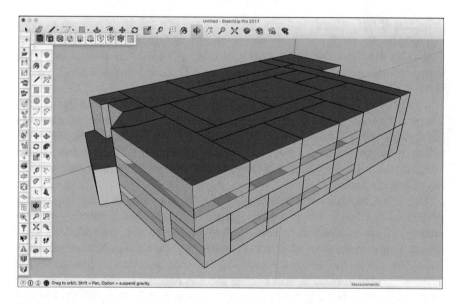

**Fig. 2.4** Building envelope drawn using the OpenStudio Plug-In for Trimble SketchUp

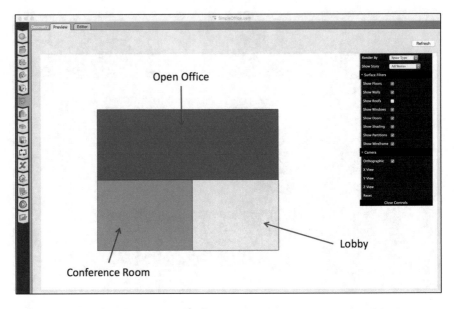

**Fig. 2.5** Simple office building with three Spaces and two Space Types on second floor

definition of typical loads and schedules for a specific activity is referred to as an OpenStudio Space Type.[3] Each Space can be assigned a single Space Type. Figure 2.5 illustrates the second floor of a small office building divided into three Spaces which reference two Space Types. Space Types will be discussed in greater detail in Chap. 3.

One or more OpenStudio Spaces may be grouped together to form an OpenStudio Thermal Zone (Fig. 2.6). Thermal Zones are the primary object of simulation in EnergyPlus and will be discussed in Chap. 4. Because Thermal Zones are comprised of Spaces, Space geometry defines the geometry of a Thermal Zone. In the simplest case, an entire building could be treated as a single Thermal Zone consisting of a single Space.

In practice, the division of Spaces and Space Types within a building is generally prescribed by intended use of a given space. On the other hand, selecting building zoning is a far trickier task in terms of both building <u>and</u> energy Model performance. Furthermore, simulation time increases with the number of Thermal Zones and space boundaries so excessive geometry becomes computationally expensive. It is recommended to make Spaces only as small as required to capture the different activities within a building while also supporting proper thermal zoning. Thermal Zones will be revisited in Chap. 4.

---

[3] It is worth noting that standards like ASHRAE 90.1 and California Title 24 specify assumptions and minimum performance requirements in the context of prescribed space types. The standardized definition of space types is not simply a useful concept, it is fundamental to the development of building codes and the projects that reference them.

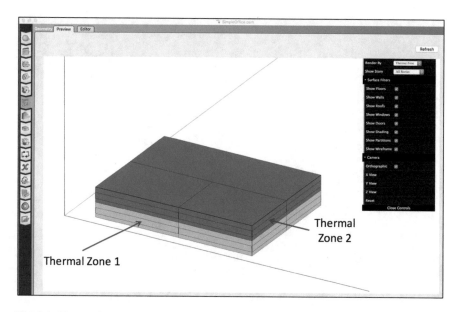

**Fig. 2.6**  Simple office building with first and second floors zoned separately

## 2.4  Surfaces

Each Space in an OpenStudio Model is bounded on all sides by a Surface. Each Surface is an infinitely thin, two-dimensional polygon that separates the volume inside a Space from the region outside the Space. The normal vector of a Surface (calculated by tracing the vertices using the right-hand rule) always points out of a Space. The collection of Surfaces associated with a Space defines the total volume of the Space. Omitting a Surface from a Space's boundary (e.g. not fully enclosing a Space) results in incorrect volume and area calculations and should be avoided.

Surfaces play a large role in defining the heat transfer into and out of Spaces and their related Thermal Zones (Fig. 2.7). Modes of heat transfer through a surface are shown in Fig. 2.8. Note that sunlight does not pass directly through the surface into the space. Transparent openings such as windows and skylights must be modeled separately as Sub-Surfaces, which are discussed later in this Chapter. Heat and mass transfer via (unintentional) infiltration is described in Chap. 3 and mechanical ventilation is introduced in Chap. 4.

Because Surfaces are infinitely thin, new modellers commonly wonder how they should describe actual building surfaces, which have non-zero thickness. The authors recommend placing Surfaces on the exterior face of all building surfaces exposed to outdoor or ground boundary conditions as shown in Fig. 2.9. Surfaces should be placed on the centreline of interior building surfaces as shown in Fig. 2.10. This technique ensures that the exterior area exposed to the outdoor environment is fully captured. Assuming that the envelope surface area is a more significant driver

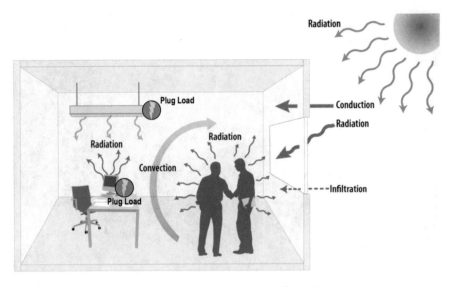

**Fig. 2.7** Heat transfer mechanisms within a space. (Credit Marjorie Schott)

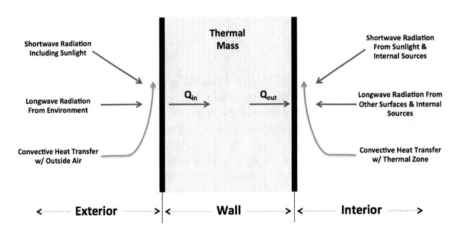

**Fig. 2.8** Interior and exterior surface heat transfer

of building thermal performance than interior heat transfer effect, this can improve thermal model fidelity. It should be noted that this approach does artificially extend floor Surfaces into wall cavities, yielding a floor area that is somewhat larger than the usable area. Additionally, the calculated air volume derived from the bounding Surfaces will include the volume of air inside the wall and ceiling cavities. For larger buildings, these effects will be minimal.

Each individual Surface may be classified as a wall, floor, or roof/ceiling. These classifications allow for calculations such as the floor area of a Space or total exterior wall area. Each Surface also specifies the outside boundary condition that is

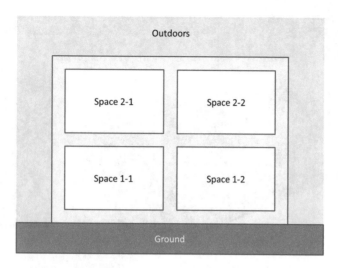

**Fig. 2.9**  Two story building with building surface thickness shown

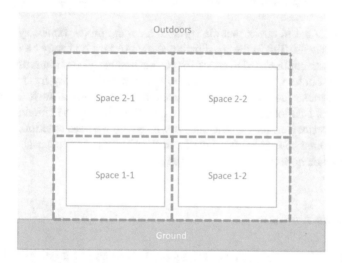

**Fig. 2.10**  Recommended placement of OpenStudio surfaces to define Spaces

applied to that Surface. Outdoor Surfaces are exposed to the exterior environment typically including outdoor air temperature and solar radiation. Ground Surfaces are in contact with the ground domain and are not typically exposed to the outdoor air temperature or solar radiation. Interior Surfaces are in contact with a Surface in another space and are exposed to the indoor conditions of the other Space instead of the outdoor environment. A special Adiabatic Surface type does not allow any heat transfer by conduction into a Space and can be useful in some advanced applications.

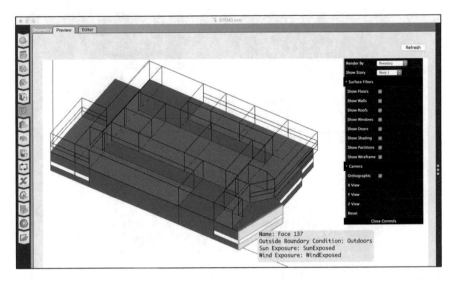

**Fig. 2.11** Inspecting geometry boundary conditions in the OpenStudio Application

It is important to check that all Surfaces have the proper boundary conditions. The Geometry (⬛) Tab in the Application shown in Fig. 2.11 may be used for this purpose. It is recommended to check for correct boundary conditions on all exterior surfaces first before hiding roof geometry and proceeding story by story to check that all interior surfaces have the correct boundary conditions. A Surface's boundary condition –not its location- dictates how it will treated by the simulation engine. Matched interior Surfaces should show up in green, whereas exteriors will appear blue and ground contact Surfaces render in light brown. Incorrect assignments will fail to simulate or produce unexpected results.

## 2.5   Constructions

Each Surface has an associated Construction. A Construction is comprised of layers of Materials as shown in Fig. 2.12. Materials are ordered from the exterior to interior Surface. Each Material layer has properties related to its heat transfer characteristics. Composite layers, such as a wood stud wall with batt insulation, are modelled with the Material properties of the overall assembly. Libraries of common construction material and assembly properties are distributed with OpenStudio. Additional Materials can also be found in the Building Component Library. Accessing materials from these sources will be discussed further in this Chapter's tutorial section. During simulation, the temperature on both sides of each Material layer is computed. Heat transfer through each Material is a function of the temperature difference across the Material, the Material's thermal resistance, and the Material's capacity to store thermal energy. This is commonly represented as an equivalent RC circuit model as shown in Fig. 2.13.

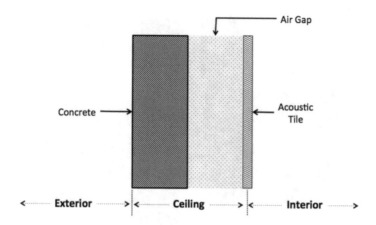

**Fig. 2.12** Material layers in a ceiling construction

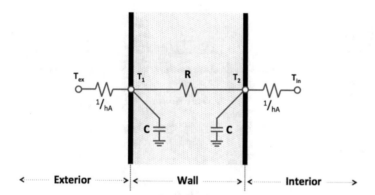

**Fig. 2.13** Equivalent RC network for a construction

Interior Surfaces between adjacent Spaces are represented by Surfaces in each Space that reference one another. By design, these adjacent Surfaces each have Constructions which mirror each other, that is their Material layers are identical but in reverse order. If a Construction is symmetric (e.g. a layer of drywall followed by wood stud wall and another layer of drywall) then both Surfaces may reference the same Construction. This concept is shown in Fig. 2.14.

## 2.6  Sub-Surfaces

Surfaces may have openings such as doors, windows, and skylights. These Sub-Surfaces are linked to a parent Surface. The Sub-Surface overlaps the parent Surface, and its area is subtracted from the parent's gross area for the purpose of heat

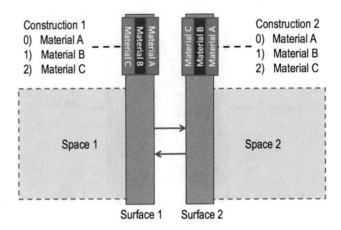

**Fig. 2.14** Mirrored constructions for adjacent surfaces

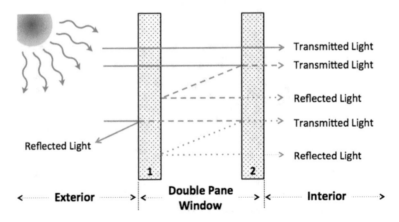

**Fig. 2.15** Heat transfer through transparent constructions

transfer. Like Surfaces, Sub-Surfaces also reference related Constructions. Constructions for opaque Sub-Surfaces such as doors may use the same opaque Materials as Surfaces. All Sub-Surfaces must be explicitly modeled. Cutting a hole in a surface or leaving a gap in the space boundary does not result in simulation of air transfer or allow light into the space. Note that solar radiation only passes through <u>transparent</u> Sub-Surfaces like windows or skylights.

Constructions for transparent Sub-Surfaces such as windows or skylights must be made of window Materials such as glass, gas filled voids, or window shading such as blinds or screens. Like opaque Materials, transparent Materials allow heat transfer through conduction as well as heat storage. However, transparent Materials also allow solar radiation to pass through or be reflected back at each layer (Fig. 2.15). Transparent properties are often provided separately for the entire solar radiation spectrum and for the spectrum of visible light.

## 2.7  Introduction to Data Inheritance in OpenStudio

At this point, many readers may be reeling at the prospect of properly assigning Space Types, Constructions, and Materials subject to the rules and constraints described in previous sections. In truth, creating correct EnergyPlus input files requires significant attention to detail, but fortunately OpenStudio comes to our rescue with a concept known as "data inheritance." Data inheritance is a key feature of object oriented software paradigms, allowing child objects to inherit data from their parents without explicit specification. For example, defining a building type allows OpenStudio to inherit a standard Space Type that is automatically applied to all Spaces within that building. For example, Spaces within a school building automatically inherit a "classroom" Space Type, the most common in schools. Explicit assignment of Space Type to any given Space becomes the exception rather than the rule, saving time and reducing the opportunity for error.

The concept of data inheritance also applies to Constructions and Materials. With OpenStudio, users may specify a top-level Construction Set that will be applied to an entire building. Each Construction Set specifies a standard Construction for a particular type of Surface or Sub-Surface (e.g. walls, roofs, windows, doors). When a Surface or Sub-Surface is queried for its Construction, the Surface first checks if a Construction has been explicitly assigned to it. If not, the Surface or Sub-Surface checks if its associated Space has a Construction set which defines a construction for it. If not, the Space's Building Story Construction Set is checked and finally the Construction Set assigned to the whole building will be checked.

The inheritance process, shown in Fig. 2.16, allows for constructions to be quickly assigned to all surfaces in the Model while also allowing explicit Construction specification for Surfaces, Spaces, or building Stories as needed.

**Fig. 2.16** Construction inheritance hierarchy

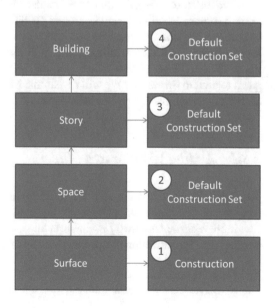

Numbers in this figure represent the relative priority of Construction or Material data that may have been explicitly assigned at any level. For example, a Construction Set that may have been explicitly assigned in a Space will take precedence over any Construction Sets that may have been assigned for the Story or Building. Of course, if the modeler has defined a specific Surface or Sub-Surface construction in a Space that will always be respected.

Typically, Construction Sets contain definitions for all types of Surfaces or Sub-Surfaces. However, Construction Sets may also include partial definitions – e.g. window Constructions, but not wall Constructions. In such cases, the inheritance search process halts as soon as it finds appropriate data. This allows, for example, a Construction Set to specify a particular window Construction to use for one Story of the building while all other Constructions would be inherited from the Building Construction Set.

Data inheritance is a key feature of OpenStudio, and understanding how it works is important for maximizing a modeler's productivity. We will reinforce this concept in Chap. 3 where Space Type inheritance will be discussed in greater detail, but inheritence applies throughout the SDK and we cannot overstate the value of understanding how it works.

## 2.8  Checkpoint One: Creating an Energy Model with an Unconditioned Envelope

In this first exercise, we will create a simple Model with a single Space and Thermal Zone. The geometry for our first Model shown in Fig. 2.17 comes from ASHRAE Standard 140 test case 600. ASHRAE Standard 140 is used to evaluate energy simulation programs, and is designed around BESTEST, a suite of building modeling test cases.[4] For this first exercise, we will only specify our simple building's envelope geometry and Constructions. This will allow us to simulate the envelope's unconditioned response to weather conditions. In Chap. 3 we will add internal heat gains related to Space use. We will revisit this Model once more in Chap. 4, adding an HVAC system for Space conditioning.

### 2.8.1  Adding Weather to a Model

To begin, we must first identify the location and weather file for our Model. We recommend that you seek out weather files associated with your own city. Weather files in EPW format are freely available at https://energyplus.net/weather. Download weather data for a location of interest. Once downloaded, examine the data using the

---

[4] https://yeungus.com/ashrae-standards-analysis-free-related-pdf.html.

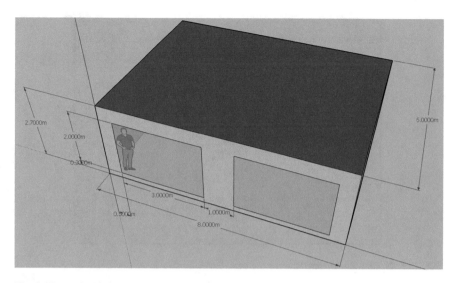

**Fig. 2.17** ASHRAE Standard 140 BESTEST Case 600 Building

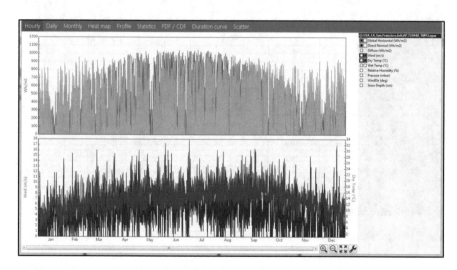

**Fig. 2.18** TMY data inspected with DView

DView Application packaged with OpenStudio (Fig. 2.18) to become acquainted with the type of information it contains. With weather files in hand, it is time to create our first OpenStudio Model.

Locate the OpenStudio Application on your computer and launch it. As the program begins running, the user is presented with a startup dialog shown in Fig. 2.19, while a new, empty Model is created. After a brief pause, the dialog is replaced with the main OpenStudio Application window shown in Fig. 2.20.

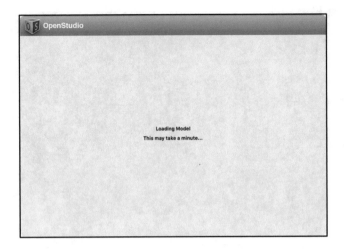

**Fig. 2.19** OpenStudio launch dialog

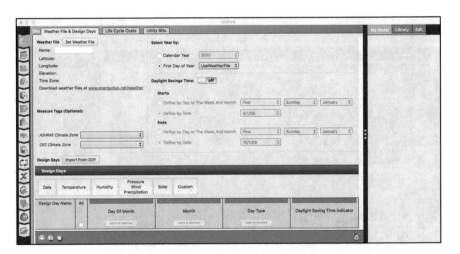

**Fig. 2.20** OpenStudio Application with a newly created model

As with any electronic document, it is a good practice to save work often and to create backups along the way. This allows you to restore your Model from the last save point rather than having to start completely over if you make a mistake. Get off to a good start by using the File Menu and selecting the Save option. A dialog opens allowing you to select the working directory and name of your first Model. Saving the empty Model creates an OSM (OpenStudio Model) file along with an identically named directory (Fig. 2.21). The OSM is an OpenStudio Model file and is the file you would send if you wanted to share your Model with someone. The corresponding Model directory will eventually contain supporting files, simulation results, and more.

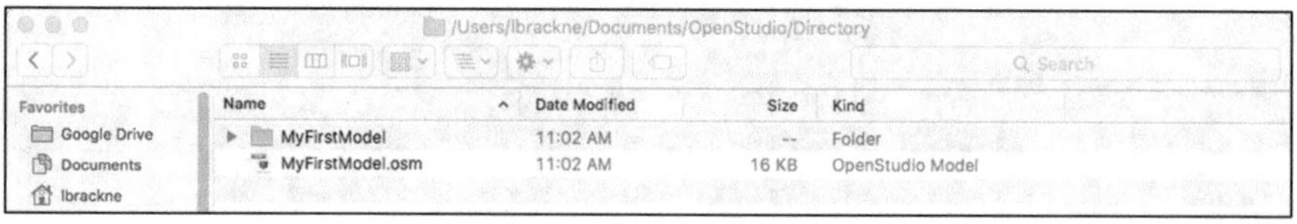

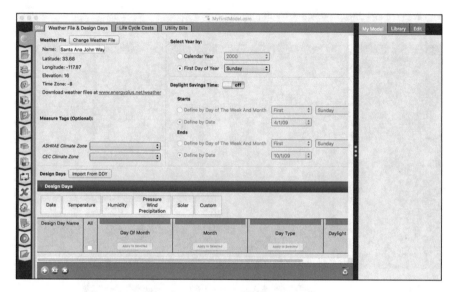

**Fig. 2.21**  A newly saved OpenStudio Model

**Fig. 2.22**  Adding EPW and DDY files in the OpenStudio Application

The OpenStudio Application initially opens in the Site (▣) Tab used to specify site details like weather. Click on the Change Weather File Button and select the EPW file you have downloaded. This will attach the weather file to your Model and import location information such as latitude, longitude, and elevation as shown in Fig. 2.22.

Save the Model once again and note that the Model directory now includes a copy of the EPW in a "files" subdirectory as shown in Fig. 2.23.

## 2.8.2  Creating New Materials

The next step in the process is to define properties for some of the Materials in our building. Click on the Constructions (▥) Tab in the Application. This Tab contains three Sub-Tabs: Construction Sets, Constructions, and Materials. We need to start by defining some basic materials so click the Materials Sub-Tab. On the left side of the Materials Sub-Tab shown in Fig. 2.24 note the categories for Material Object types including "Materials," "No Mass Materials", and so on.

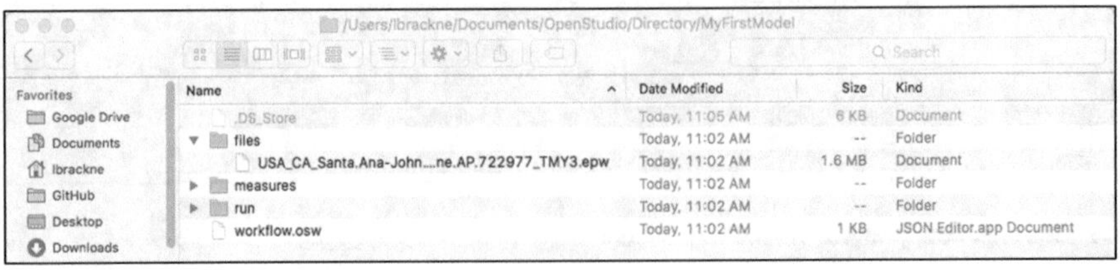

**Fig. 2.23** Model directory with EPW file

**Fig. 2.24** The OpenStudio Application Constructions Tab and Materials Sub-Tab

These are the different Material types supported in OpenStudio, which closely mirrors materials supported by EnergyPlus. Detailed information about these material types can be found in the EnergyPlus I/O Reference and Engineering Guides.[5] A mapping between Material types and EnergyPlus materials is given in Table 2.1.

We will start by making a new Material of type "No Mass Materials." Click on that label and press the ⬜ Button. This will create a new Object as shown in Fig. 2.25.

Name the Object "R25 Insulation" and set the properties as shown in Table 2.2. Repeat this process for the remaining material objects in the Table.

### 2.8.3  Importing Materials from Existing Models

We have now made several Material objects by hand and have a better understanding of how Materials are represented in OpenStudio and EnergyPlus. However, this process is tedious and prone to error when performed frequently. Instead, let's

---

[5]The authors recommend bookmarking this site, because we will refer to it frequently in later Chapters – https://energyplus.net/documentation.

**Table 2.1** OpenStudio to EnergyPlus Material type mapping

| OpenStudio Material type | EnergyPlus Material type |
|---|---|
| Materials | Material |
| No Mass Materials | Material:NoMass |
| Air Gap Materials | Material:AirGap |
| Simple glazing system window Materials | WindowMaterial:SimpleGlazingSystem |
| Glazing window Materials | WindowMaterial:Glazing |
| Gas window Materials | WindowMaterial:Gas |
| Gas mixture window Materials | WindowMaterial:GasMixture |
| Blind window Materials | WindowMaterial:Blind |
| Screen window Materials | WindowMaterial:Screen |
| Shade window Materials | WindowMaterial:Shade |
| Air wall Materials | N/A |
| Infrared transparent Materials | Material:InfraredTransparent |
| Roof vegetation Materials | Material:RoofVegetation |
| Refraction extinction method glazing window Materials | WindowMaterial:Glazing:RefractionExtinctionMethod |
| Glazing group thermochromic window Materials | WindowMaterial:GlazingGroup:Thermochromic |

explore how to load Materials from OpenStudio's built-in libraries as well as the online Building Component Library. The OpenStudio Application allows a user to extract data like Space Types, Constructions, and Materials from any OSM by loading it as a "Library". Users can then incorporate pieces of information from the Library OSM into their current working Model.

Select Load Library from the File Menu. This will bring up a dialog where you can browse for other OSM files to use as a Library. The dialog defaults to a directory within the OpenStudio installation package containing a number of useful OSMs as shown in Fig. 2.26, but the user is free to navigate to other directories containing OSMs.

For the purpose of this exercise, select CECTemplate.osm. On the right side of the Materials Tab click on the ▢Library Tab and expand the Materials section just beneath it. You can now see all of the Material objects in the Library. Any of these can be added to the current Model by right clicking, dragging, and dropping them on the space labeled "Drag From Library." Once a Material has been added to the Model it may be inspected, edited, and used in the same manner as any hand-generated

**Table 2.2** Material properties for our first Model

| Properties | Material name | R25 Insulation | Air gap | Plasterboard | Fiberglass quilt | Roof deck | Timber flooring | Door shell | Interior wall | Wood siding | Glass | Window air fill |
|---|---|---|---|---|---|---|---|---|---|---|---|---|
| | Material type | No mass | Air gap | Materials | Materials | Materials | Materials | Materials | Materials | Materials | Glazing window | Gas window |
| Material properties | Roughness | Rough | – | Rough | Rough | Rough | Rough | Very smooth | Rough | Rough | – | – |
| | Thickness (m) | – | – | 0.012 | 0.066 | 0.019 | 0.025 | 0.0032 | 0.2 | 0.009 | 0.0032 | 0.013 |
| | Conductivity (W/m-K) | – | – | 0.16 | 0.04 | 0.14 | 0.14 | 1.06 | 0.51 | 0.14 | 1.06 | – |
| | Resistance ($m^2$-K/W) | 25.075 | 0.1588 | – | – | – | – | – | – | – | – | – |
| | Density (kg/$m^3$) | – | – | 950 | 12 | 530 | 650 | 2500 | 1400 | 530 | – | – |
| | Specific heat (J/kg-K) | – | – | 840 | 840 | 900 | 1200 | 750 | 1000 | 900 | – | – |
| | Thermal Absorptance | 0.9 | – | 0.9 | 0.9 | 0.9 | 0.9 | 0.9 | 0.9 | 0.9 | 0.9 | – |
| | Solar Absorptance | 0.6 | – | 0.6 | 0.6 | 0.6 | 0.6 | 0.6 | 0.6 | 0.6 | 0.6 | – |
| | Visible Absorptance | 0.6 | – | 0.6 | 0.6 | 0.6 | 0.6 | 0.6 | 0.6 | 0.6 | 0.6 | – |
| | Optical data type | – | – | – | – | – | – | – | – | – | Spectral average | – |
| | Solar transmittance at normal incidence | – | – | – | – | – | – | – | – | – | 0.862 | – |
| | Front solar reflectance at normal incidence | – | – | – | – | – | – | – | – | – | 0.078 | – |

| | | | | | | | | | | | | | |
|---|---|---|---|---|---|---|---|---|---|---|---|---|---|
| Back solar reflectance at normal incidence | – | – | – | – | – | – | – | – | – | – | – | – | 0.078 | – |
| Visible transmittance at normal incidence | – | – | – | – | – | – | – | – | – | – | – | – | 0.913 | – |
| Front visible reflectance at normal incidence | – | – | – | – | – | – | – | – | – | – | – | – | 0.082 | – |
| Back visible reflectance at normal incidence | – | – | – | – | – | – | – | – | – | – | – | – | 0.082 | – |
| Infrared transmittance at normal incidence | – | – | – | – | – | – | – | – | – | – | – | – | 0 | – |
| Front infrared emissivity | – | – | – | – | – | – | – | – | – | – | – | – | 0.84 | – |
| Rear infrared emissivity | – | – | – | – | – | – | – | – | – | – | – | – | 0.84 | – |
| Gas type | – | – | – | – | – | – | – | – | – | – | – | – | – | Air |

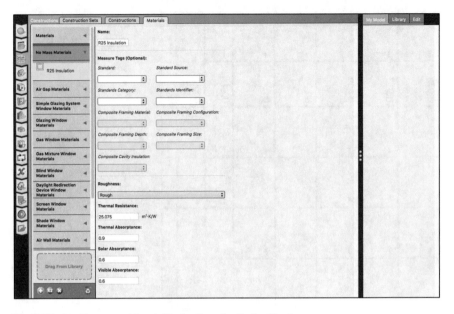

**Fig. 2.25**  Creating a new Material in the OpenStudio Application

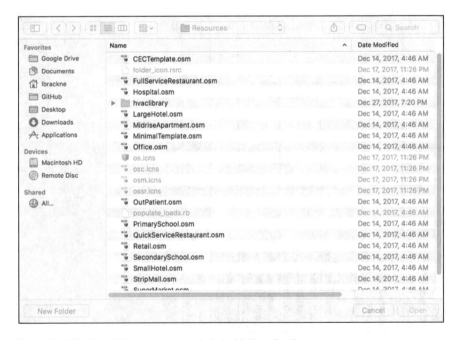

**Fig. 2.26**  OSM Load Library options included with OpenStudio

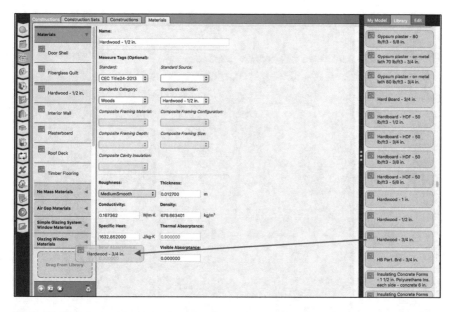

**Fig. 2.27** Dragging and Dropping Materials from Library into a Model

Material. Figure 2.27 illustrates inspecting the properties of a ½″ Hardwood Material while adding a ¾″ Hardwood Material from the Library. We won't be using any Materials from this Library for our example Model, but feel free to experiment with them.

### 2.8.4   Importing Materials from the Building Component Library

Besides existing OSMs, another source for Material data is the Building Component Library (BCL). The BCL is an online repository of energy modeling tool agnostic data, but it has been designed to integrate with OpenStudio tools particularly well. A web interface to the BCL is available at http://bcl.nrel.gov, however it may also be accessed within the Application under the "Components & Measures" Menu category. Prior to using the BCL from within the Application, you will need to create an account on the BCL web page, navigate to "My Dashboard," and make note of your unique BCL "API Key." The OpenStudio Application will request this Key the first time you attempt to access BCL content.

Selecting "Find Components" within the Application launches an integrated browser that allows the user to search, inspect, and download components from the BCL directly into your Model. Click on the Material category and the Opaque

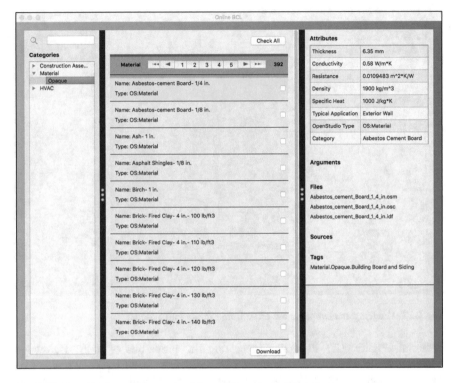

**Fig. 2.28**  Downloading a Construction Material from the Building Component Library

sub-category to bring up the list of materials shown in Fig. 2.28. Select one or more of these objects and press the [Download] Button at the bottom of the window to download Materials and add them to your Model. We will revisit the BCL in Chap. 6.

## 2.8.5  Creating Construction Assemblies

Now that Materials have been added to the Model, it is time to combine them to create Constructions using the [Constructions] Sub-Tab (Fig. 2.29). Select the "Constructions" Object type and press the ⊡ Button to create a new Construction Object. As in the [Materials] Sub-Tab, the right-hand pane may be used to inspect objects currently loaded in the working Model or library. Click either [My Model] or [Library] and select Materials to drag onto the "Drag from Library" drop zone in the Construction editor. This adds the selected Material to the currently edited Construction. Materials may be deleted from a Construction by clicking the ⊗ Button next to it. Remember that Construction layers are ordered from the outside layer at the top and move towards the Space interior as you progress down the layer list in the editor.

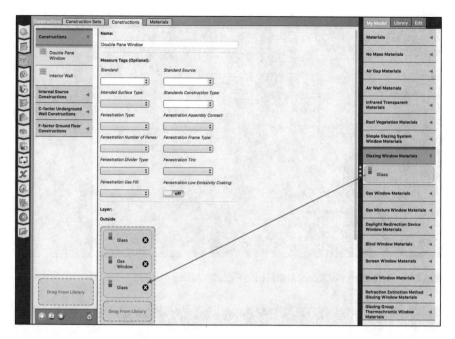

**Fig. 2.29**  Creating a double pane window Construction in the OpenStudio Application

**Table 2.3**  Defining Construction layers

| Object name | | Interior wall | Double pane window | Door | Floor | Roof | Wall |
|---|---|---|---|---|---|---|---|
| Space exterior | **Layer 1** | Interior wall | Glass | Door shell | R25 insulation | Roof deck | Wood siding |
| | **Layer 2** | – | Window air fill | Air gap | Timber flooring | Fiberglass quilt | Fiberglass quilt |
| Space interior | **Layer 3** | – | Glass | Door shell | – | Plasterboard | Plasterboard |

Use the Constructions Sub-Tab to create the following Construction objects listed in Table 2.3.

Note that Libraries may also contain complete Constructions. Selecting the Constructions category under ▭ Library lists any Constructions that may have been imported from an OSM. As with Materials, Library Constructions may be dragged and dropped into the working Model to save time and reduce the opportunity for error. Adding Constructions in this manner also adds all requisite Materials from the Library with no additional effort required.

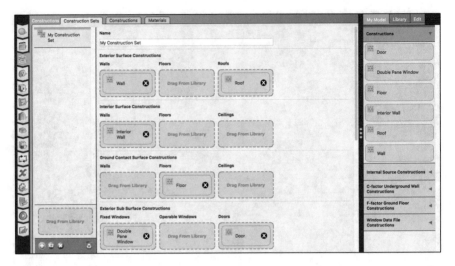

**Fig. 2.30** Building a Construction Set in the OpenStudio Application

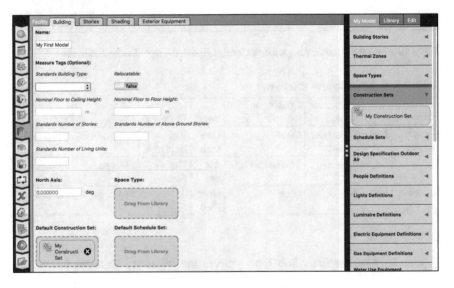

**Fig. 2.31** Setting a Construction Set as the Default for a building

## 2.8.6  Creating a Construction Set

The next step in our process is to combine individual Constructions into a Construction Set that may be applied to the entire building. Go to the [Construction Sets] Sub-Tab and create a new Construction Set named "My Construction Set" using the 🔲 Button. Drag constructions from [My Model] into the appropriate slots within the Construction Set as shown in Fig. 2.30.

The last step is setting a Construction Set as the default for our entire building is performed on the Building (■) Tab. Select "My Construction Set" and drag it to the Default Construction Set slot as shown in Fig. 2.31.

As with Materials and Constructions, entire Construction Sets may be pulled in from Library. This allows the modeler to quickly assign Constructions to walls, floors, roofs, etc. at the Building level with a single drag-and-drop operation. Using pre-built Construction Sets is the most common way to begin creating a new Model in OpenStudio, but it is important for us to understand how Construction Sets are formed so that we may evaluate their suitability and modify them as needed for any given project.

### 2.8.7 Creating Geometry with the OpenStudio Floor Plan Editor

Having assigned weather files and a Construction Set to our Model, we are now ready to define the building's geometry. Click on the Geometry (■) Tab to open the OpenStudio Application's Geometry editor and previewer. You will initially see an empty window on the Preview Sub-Tab because the current Model does not yet contain any geometry. Click on the Editor Sub-Tab to reveal OpenStudio's floor plan geometry editor. The editor initially presents the user with some introductory text and the option to create a new floor plan. Clicking the New Button opens the QuickStart dialog shown in Fig. 2.32. Select "New" to create a new floor plan on an empty grid (Fig. 2.33).

Figure 2.33 points out key sections of the floor plan editor. These include:

- Drawing Editor Sub-Tabs used to:

  Draw the floor plan
  Assign Construction Sets, Space Types, Thermal Zones, etc. to regions of the floor plan
  Assign windows, doors, etc. to Surfaces on the floor plan

**Fig. 2.32** Floor plan geometry Editor Quick Start Dialog

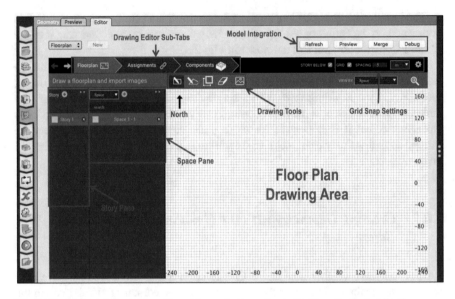

**Fig. 2.33**  Annotated floor plan editor with empty floor plan in the OpenStudio Application

- Drawing Tools including:

  Rectangular drawing tool
  Polygon drawing tool
  Floor-to-floor copy tool
  Rectangular erase tool
  Background image import for floor plan tracing

- Grid Settings used to manage draw snapping behavior
- Story pane allowing addition and editing of stories to the Model
- Space pane primarily used to add, delete, and edit Spaces from the Model

  The choice menu near the top of this pane also enables editing of shading surfaces and background images

- Model integration Buttons that convert the 2D floor plan drawing into OpenStudio 3D Model geometry

For this simple Model, we will rely on the editor's "grid snap" feature, which is active when the Grid check box is selected near the upper right-hand corner of the window. Grid spacing is set immediately to right of the check box. For this example, specify a 2 m grid and use a mouse wheel or track pad to zoom in so that an 8 m × 6 m rectangle will fit in the drawing area.

Notice that the editor opened with a single Story and Space already created for us. They are named "Story 1" and "Space 1-1" respectively. All that remains is to select the ▣ drawing tool to create an 8 m × 6 m rectangle for "Space 1-1" as shown in Fig. 2.34. If you make a mistake, use the undo/redo (◀▶) Buttons near the upper

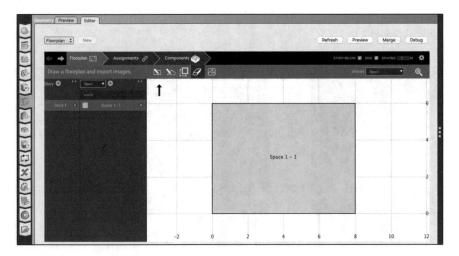

**Fig. 2.34** Drawing BESTEST Case 600 in the OpenStudio Application

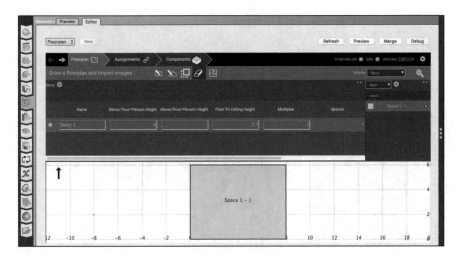

**Fig. 2.35** Editing floor height in the OpenStudio Application

left-hand corner of the window. Drawing adjacent rectangles will add to the currently selected Space, while the ⬛ tool can be used to remove rectangular regions.

The height of our single Space is defined within the Story that contains it. Stories may be added and modified at the Stories pane of the editor window. Click on the ⬛ Button to expand the Story editor as shown in Fig. 2.35, and change the floor to ceiling height from its default value to 2.7 m.

Near the top of the editor, click on the Assignments Sub-Tab. This Sub-Tab allows us to assign Construction Sets, Thermal Zones, Space Types, and more to sections of the Model. By default, the Assignment choice menu lists "Thermal

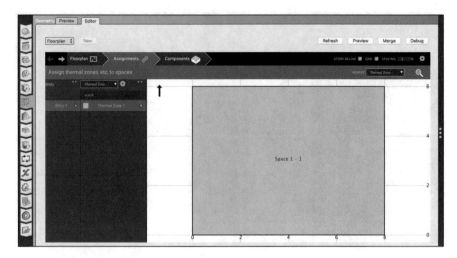

**Fig. 2.36**  Model with "Space 1-1" assigned to "Thermal Zone 1"

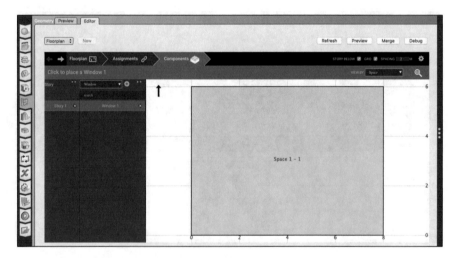

**Fig. 2.37**  Adding a Window definition with the Geometry editor

Zones" on the left-hand side of the window. Our Model doesn't yet have any, so press the ⊙ Button immediately to the right of the choice menu to create one, then click on Space 1-1 to assign it to that Thermal Zone as shown in Fig. 2.36.

Finally, we must add the two windows shown in Fig. 2.17 using the Components Sub-Tab. In the Components Sub-Tab, the Spaces Pane changes, allowing us to add Window definitions to the model (Fig. 2.37). Click the ⊙ Button to add a new definition, then expand it with the adjacent ▦ Button.

Define "Window 1" to have width of 3 m, height of 2 m, and sill height of 0.2 m. Mousing over the perimeter walls indicates locations where windows will fit. Place windows on the South façade as shown in Fig. 2.38.

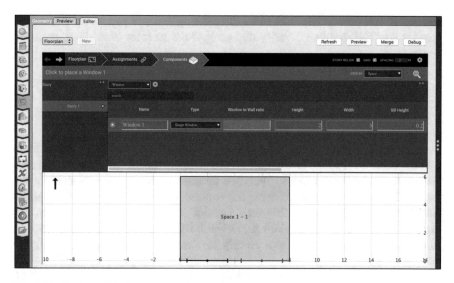

**Fig. 2.38** Placing a Window with the Geometry editor

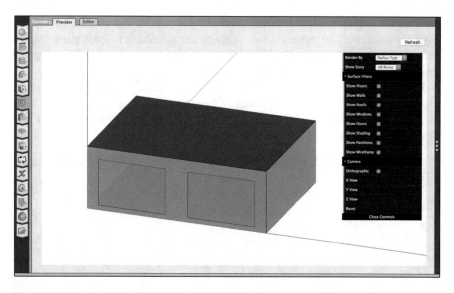

**Fig. 2.39** Inspecting the BESTEST Case 600 Building in the OpenStudio Application

We are now done defining the geometry for our first Model. Press the Merge to translate your 2D floor plan to a full 3D model usable by OpenStudio. Switching back to the Preview Sub-Tab should look similar to Fig. 2.39. If your Model doesn't look quite right, switch back to the editor, make changes, and re-merge your work until you are satisfied.

In this view, the green axis is pointing North. Next to "Render By" choose "Boundary", this will color all of your surfaces by their boundary conditions.

Check all of the boundary conditions by clicking on each surface. Next, choose Render By "Construction". Verify that each of the surfaces have the appropriate construction assigned.

### 2.8.8  Running Your First Energy Simulation

You are almost ready to run your first simulation. Before doing so, we want to configure the simulation to write out some detailed time series data to examine. Select the Variables (⧆) Tab to specify a number of detailed results that we can request from our simulation. Toggle the request Button to "On" next to "Surface Inside Face Temperature," "Surface Outside Face Temperature," "Zone Air Temperature," and "Zone Outdoor Air Drybulb Temperature." Also change the logging rate for each variable from Hourly to Timestep to capture data more frequently as shown in Fig. 2.40.

We are now ready to run the simulation. Go to the Run (▣) Tab and press the ⊙ Button (Fig. 2.41). You will see simulation output until the EnergyPlus simulation completes.

Once the simulation is complete, select the Reports (▣) Tab to view simulation results. By default, two reports are available from the selector in the upper left-hand corner of the window: OpenStudio Results and EnergyPlus Results. The standard OpenStudio Results report is shown in Fig. 2.42.

Related to this Chapter's focus on building envelope, clicking on that report heading takes us directly to a summary of high level information about the building's shell as shown in Fig. 2.43. This can be a useful check to ensure that Constructions were applied as expected.

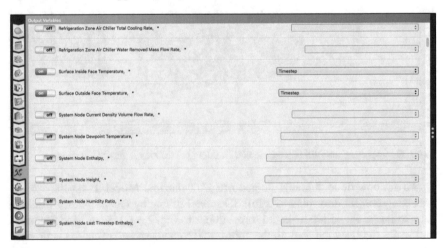

**Fig. 2.40**  The OpenStudio Application Output Variables Tab

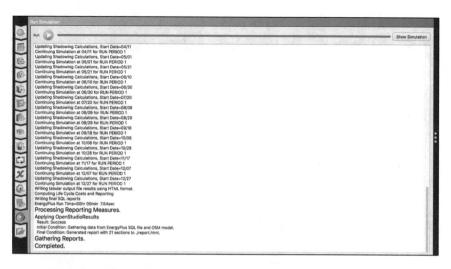

**Fig. 2.41**  The OpenStudio Application Simulation Run Tab

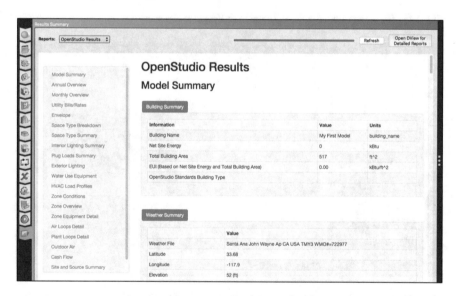

**Fig. 2.42**  The OpenStudio Standard Report within the Results Summary Tab

Figure 2.44 highlights another part of the summary report, presenting our building's Zone temperature and humidity for the entire year that we simulated. Since the air in our small building was unconditioned and subject to the ambient conditions described by our weather file, a great many hours spent throughout the year in our tiny building would be considered uncomfortable. This is why we use mechanical systems to condition spaces in our buildings and will be the subject of Chap. 4.

**Fig. 2.43**  The OpenStudio Standard Report with Envelope Summary

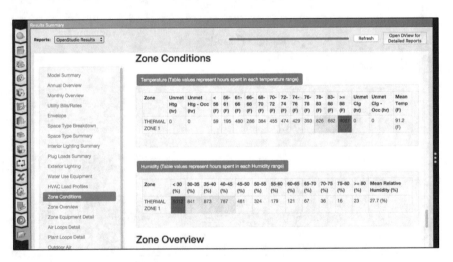

**Fig. 2.44**  The OpenStudio Standard Report with Zone Condition Summary

Changing the report selector near the top of the window switches to an alternative report. This summary (Fig. 2.45) is produced by the EnergyPlus engine itself and can be a useful supplement to the higher-level information presented in the OpenStudio report.

Since this represents our first complete simulation, it's worth revisiting the Model directory structure we discussed back in Sect. 2.8.1. Recall that the Application saves the OSM along with a directory structure that it populates throughout the

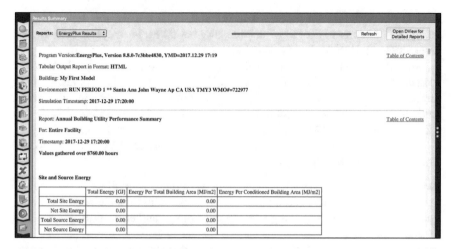

**Fig. 2.45** The EnergyPlus Report Viewed within the Results Summary Tab

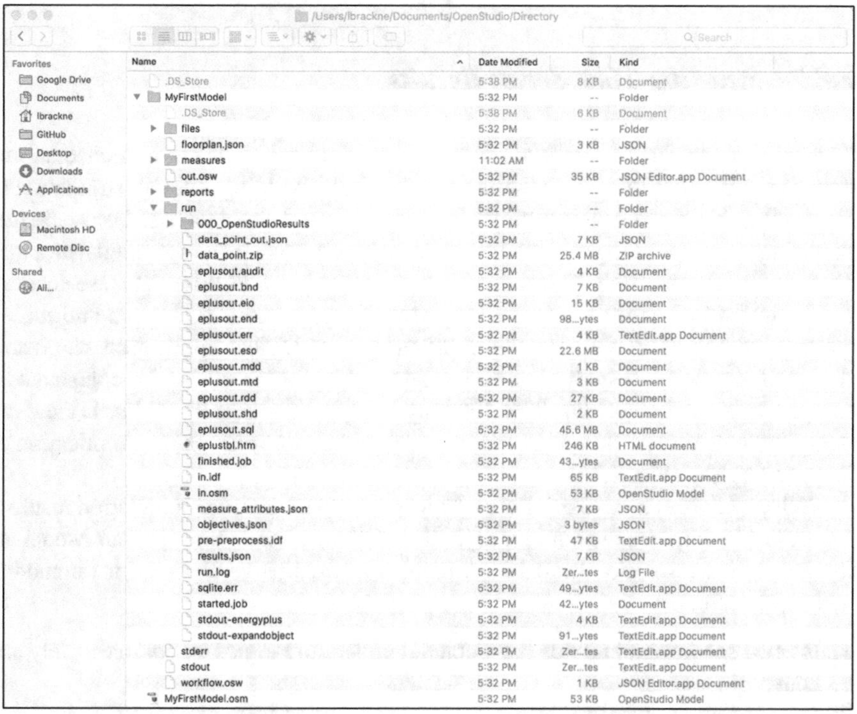

**Fig. 2.46** Model directory after simulation

modeling and simulation process. We first noted that our weather EPW file was neatly tucked away in the "files" subdirectory but take a look in Fig. 2.46 at what OpenStudio produced when the simulation ran.

We will discuss a few of these files in subsequent Chapters, but let's take a moment to point out some of the more significant subdirectories and files. We have already seen the contents of the "reports" subdirectory in the Application itself. Both the EnergyPlus and OpenStudio reports are located here as simple HyperText Markup Language (HTML) documents viewable in any web browser.

Within the run directory are a number of files generated by OpenStudio and EnergyPlus itself. Key files of note include:

**in.osm** – The final[6] OpenStudio Model prior to calling EnergyPlus

**in.idf** – The EnergyPlus input file created by OpenStudio for simulation

**eplusout.err** – A file containing simulation warnings and error messages used for troubleshooting

**eplusout.sql** – A time series database of simulation results used by plotting software like DView

### 2.8.9   Studying Time Series Results

The eplusout.sql file is a natural segue into the final topic for our first checkpoint exercise - investigating time series results from building energy simulations. EnergyPlus is able to generate very large sets of time series data based on standard reporting variables along with additional variables we may have specified using the OpenStudio Application. These data sets are stored in a SQLite database format (SQL) instead of comma space delimited text files to save disk space and make subsequent access of the data more efficient. The DView application has been designed expressly to study time series data and is able to utilize the eplusout.sql file generate by EnergyPlus. This file may be opened manually within DView or using the convenient Button in the upper right-hand corner of the Application's Reports (■) Tab (Fig. 2.45).

DView provides a number of ways to inspect and explore our simulation results. Figure 2.47 compares the outdoor drybulb and Zone temperatures for our Model at the finest time resolution available. A number of qualitative observations immediately spring forth:

1. Interior space temperatures lag ambient temperature trends by approximately an hour;
2. Interior temperature profiles are "smoother," having been effectively dampened by the inherent thermal capacitance of the structure;

---

[6]Final model? I thought I was working on the final model in the OpenStudio Application! The distinction between models you edit in the OpenStudio Application and the "final" model that is actually simulated will become clearer in Chap. 6.

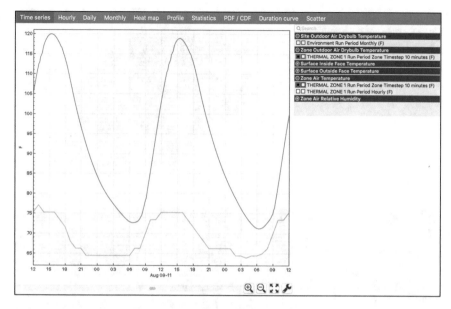

**Fig. 2.47** Comparison of Outdoor and Zone temperatures in DView

3. Interior temperatures are elevated above ambient by radiant energy admitted through the South-facing fenestration;
4. And, lastly, this building would be an incredibly unpleasant structure to live or work in!

DView provides other ways to explore simulation results – for example examining exterior or interior surface temperatures. As an example, consider the temperatures of the South-facing façade of our building. Using the Geometry (■) Tab previewer, we can inspect the South façade and note that OpenStudio has automatically assigned it the name "Face 2" (Fig. 2.48).

Once identified, variables associated with that surface's name can be selected in DView to plot surface temperatures in a variety of useful ways. Figure 2.49 displays the exterior surface temperature of the South façade as a "heatmap," a colored chart that allows the user to quickly identify thermal extremes by month of the year and time of day.

Figure 2.49 displays monthly "profiles" for the exterior and interior surface temperatures of the South Façade by averaging daily temperatures for each day of every month. Both visualizations can be useful in spotting performance trends, aberrant behavior in simulations, etc. We will make greater use of DView in Chaps. 3, 4, and 5 to examine Space, Zone, and HVAC system behavior in detail.

**Fig. 2.48** Identifying the
Surface name for the South
Facing Façade

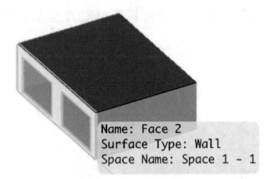

Name: Face 2
Surface Type: Wall
Space Name: Space 1 - 1

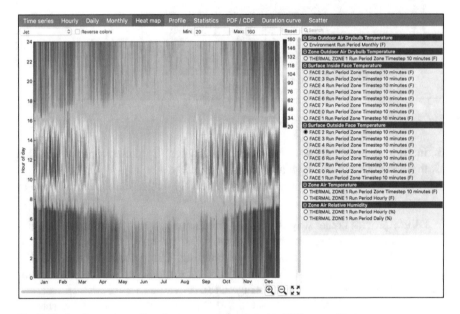

**Fig. 2.49** South exterior wall surface temperature plotted in DView as a Heatmap

## 2.9   Checkpoint Two: Energy Model of a School

Now we are ready to begin work on our real capstone Model that we will improve
upon throughout this book. We will be attempting to create a simple primary school
Model using pre-built Library data to speed up the process. A floor plan for this
building is shown in Fig. 2.51. Use the following steps to get started:

1. Create a new OpenStudio Model.
2. Under the Preferences Menu make sure that the default units are set to English
   (I-P).
3. Navigate to the Site (■) Tab and:

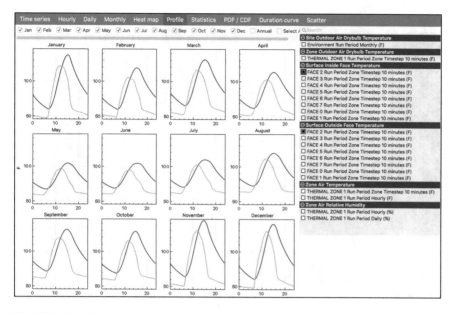

**Fig. 2.50** Monthly temperature profiles for Exterior and Interior Surfaces on South Façade

    (a)  Select the *USA_CO_Golden_NREL.72466_TMY3.epw* weather file.
    (b)  Specify ASHRAE Climate Zone 5B from the Optional Measure Tags menu.[7]

4. Pick "Load Library" from the File menu and import *PrimarySchool.osm* to load the Model library with Construction Sets and other appropriate data.
5. Navigate to the Constructions (▦) Tab, select the Construction Set named *90.1–2010-CZ5–6-PriSchl*, and drag it into your Model as shown in Fig. 2.50.
6. Navigate to the Buildings (▮) Tab, select the Construction Set you just added, and drag it to become the Default Building Construction Set.
7. Save your Model.

Now we're ready to define our Envelope and Spaces using the Geometry (▮) Tab. Create a new floor plan with the editor. Once again, decline to place this Model on a map. Select an appropriate grid spacing that will allow you to comfortably draw the floor plan shown in Fig. 2.51. Add additional Spaces with the ◉ Button and use the ◣ Tool to add them to the floor plan. The drawing cursor will turn into a larger red dot when you are on top of an adjacent vertex, which can help avoid leaving gaps between Spaces.

Expand the Space pane with the ▦ Button to change Space names and colors to your liking. Your finished floor plan should look like the one shown in Fig. 2.52.

To complete this stage of your primary school Model:

---

[7]Will discuss the significance of this in a later chapter, but for now know that this choice helps OpenStudio select Constructions and HVAC systems that are appropriate for Climate Zone 5B where our school is located.

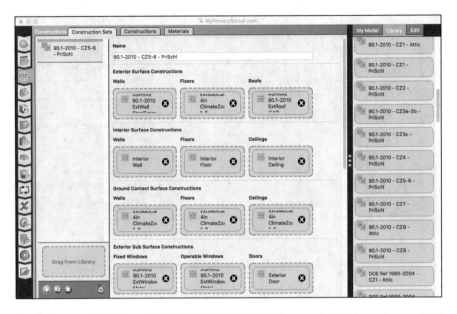

**Fig. 2.51** Empty Model Loaded with ASHRAE 90.1-2010 Construction Set for a primary school

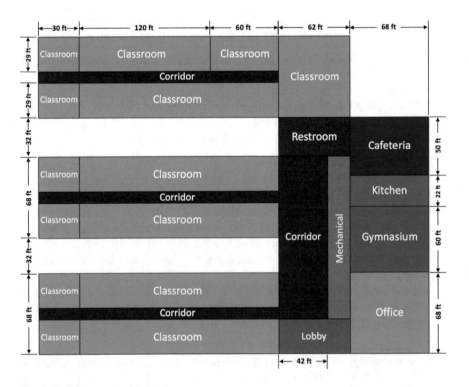

**Fig. 2.52** Primary school floor plan

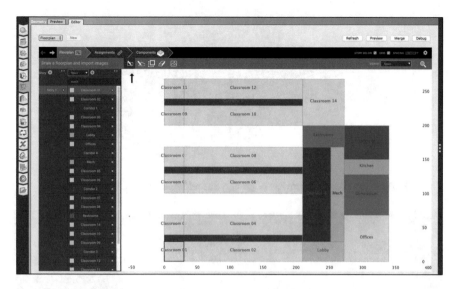

**Fig. 2.53** Primary school floor plan drawn with the Geometry editor

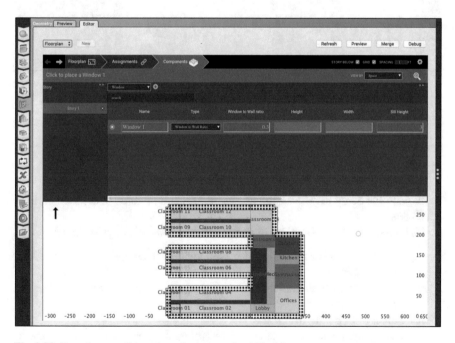

**Fig. 2.54** Boundary condition view of primary school Model

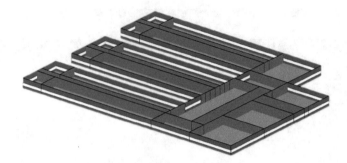

**Fig. 2.55** Boundary condition preview of primary school Model in editor

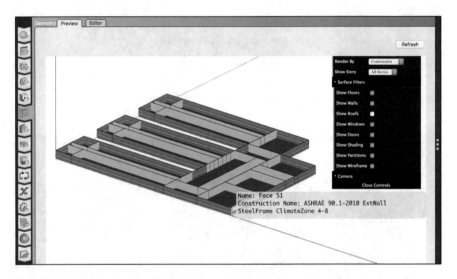

**Fig. 2.56** Construction View of primary school Model

1. Expand the Story pane with the ▣ Button to specify a floor to ceiling height of 13 feet.
2. Use the Components Sub-Tab to define and add windows to your model.

   (a) Select a "Window to Wall Ratio" type window to create a band of windows on each surface.
   (b) A Window to Wall Ratio of 0.3 will produce a window roughly 4 feet tall (0.3 × 13 feet)
   (c) Select a sill height of 3 feet.
   (d) Add a window to each Space Surface around the perimeter of your Model as shown in Fig. 2.53.

3. Use the Assignments Sub-Tab to create Thermal Zones for each Space in the model. Change their names and colors to your liking.

**Fig. 2.57** Unconditioned primary school envelope summary report

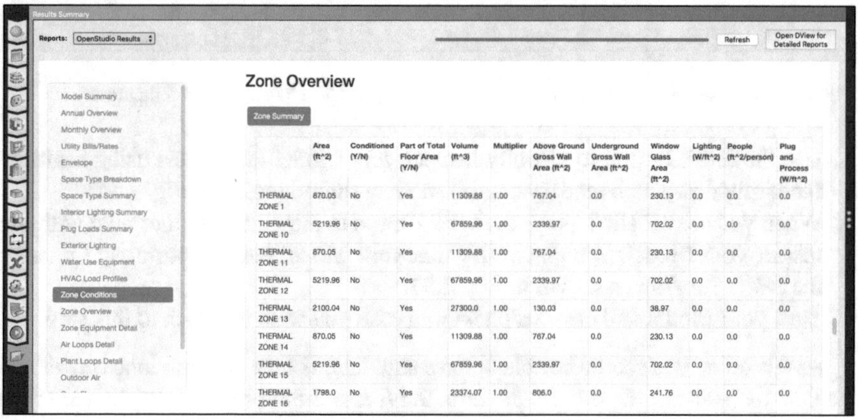

**Fig. 2.58** Unconditioned primary school zone overview

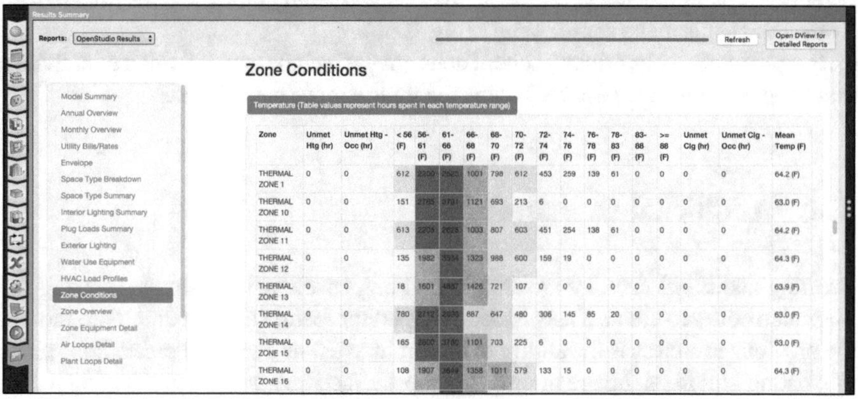

**Fig. 2.59** Unconditioned primary school zone conditions

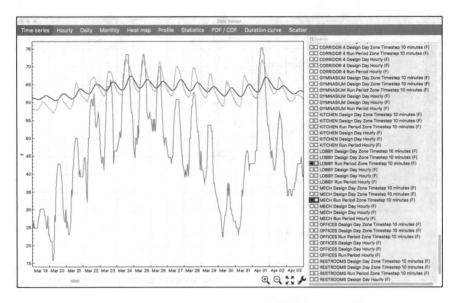

**Fig. 2.60** DView time series plots of Outdoor, Lobby, and Mechanical room temperatures

4. Use the ⟨Preview⟩ Button to visually inspect your Model for defects using multiple render modes. The boundary condition view should look like Fig. 2.54.
5. When you are satisfied, press the ⟨Merge⟩ Button to integrate the floor plan with the rest of your Model. You can verify that your Model merged correctly by using the ⟨Preview⟩ Sub-Tab as shown in Fig. 2.55.
6. Save your Model and create a backup in case you wish to return to this step.

At this point, you should be able to run a simulation of the unconditioned envelope. Review the standard reports (Figs. 2.56, 2.57, and 2.58) to see if they make sense given what you have learned so far.

Consider the time series plots shown in Fig. 2.59 that compare the outdoor temperature (dark green) with the Lobby (red) and Mechanical Room (blue) Zone temperatures. The Lobby temperature fluctuates significantly more than the Mechanical Room. Does this make sense? Why? (Fig. 2.60).

Remember that our primary school does not yet include any definition of activities that happen within each Space. That is the topic for our next Chapter.

## 2.10  Additional Exercises

Creating additional exercises related to building envelope is as simple as using the floor plan editor to create a new Model. One editor feature that we didn't mention in the previous exercises is the ability to import images for tracing to create floor plans of existing buildings. Image imports are most effectively used when creating a new

**Fig. 2.61**  Geolocating a new Model with the floor plan editor

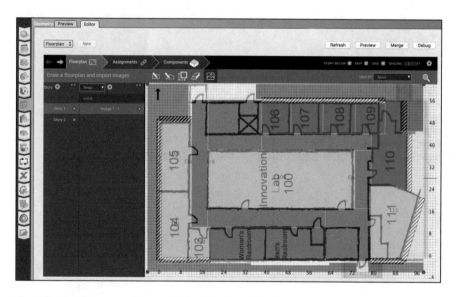

**Fig. 2.62**  Tracing over an image with the floor plan editor

floor plan and selecting "New with Map" from the floor plan quick start dialog shown in Fig. 2.32. The interface allows the user to search for a specific address, pan, zoom, and rotate to establish a region for drawing the foot print of a proposed or existing building (Fig. 2.61). The ▨ Button is then used to select an image that can be scaled, rotated, and placed on top of the map for tracing spaces. Figure 2.62

shows a snap shot of a fire escape drawing that was used to create the first floor of the Model shown in Figs. 2.3 and 2.4. Multiple images may be imported and associated with various floors of the Model.

As an exercise:

- Take a photo of a fire escape drawing of the building you study or work in,
- Create a new OpenStudio model with an appropriate weather file,
- Enter the building's address in the floor plan editor to geolocate it,
- Import your fire escape photo(s) - scaling and rotating as needed to register it on the map,
- Trace Spaces on your Model using the imported image(s),
- Import an appropriate Library of Constructions from a similar Model (e.g. School, Office, etc.),
- Assign Thermal Zones to your Spaces so you can simulate the unconditioned response, and
- Save your Model for use in subsequent Chapters.

Compare your Model with the previous exercises, when run with different weather files. Do your results make sense?

## References

Wilcox S, Marion W (2008) User's Manual for TMY3 Data Sets, NREL/TP-581-43156, National Renewable Energy Laboratory, Apr 2008
ANSI/ASHRAE Standard 169-2013, Climatic data for building design standards, 2013
http://bcl.nrel.gov
https://energyplus.net/documentation
https://energyplus.net/weather
https://yeungus.com/ashrae-standards-analysis-free-related-pdf.html

# Chapter 3
# Defining Energy Uses and Spaces

## 3.1 Energy Uses and Thermal Loads

In Chap. 2 we defined the building envelope, the ambient weather conditions it is exposed to, and the interior Spaces that a building is subdivided into. Of course, the activities that take place in those Spaces are significant drivers for energy consumption as well as the reason buildings exist in the first place. In this Chapter, we will gain a better understanding of how Space occupancy and energy end uses are defined by OpenStudio. As with Constructions, the amount of data required to fully specify Space loads is significant, and we will come to appreciate how OpenStudio Libraries and data inheritance make this process both fast and consistent.

Building end uses may consume energy directly as is the case with lighting, electric, and gas Equipment; however, these end uses may also add heat to the Spaces in which they are contained. This heat may impact the heating or cooling energy, which must be provided by the building's HVAC systems. Modeling these types of interactions is an important feature of whole building energy simulation. Another significant source of thermal loading within Spaces are the occupants themselves; people contribute both sensible and latent heat through physical activity, perspiration, and respiration. Infiltration, unconditioned air that leaks into Space, is also considered a load that will be discussed in this Chapter. Lastly, while they don't generate heat or consume energy explicitly, we will also consider the role that the thermal mass of inanimate objects within Spaces plays in storing and releasing thermal energy.

It is important to note that Space loads are a strong function of occupant behavior. As such, this step in building energy modeling is arguably the most subjective and error prone part of the process. Whereas the thermal properties of

The original version of this chapter was revised. A correction to this chapter can be found at https://doi.org/10.1007/978-3-319-77809-9_10

**Electronic Supplementary Material:** The online version of this chapter (https://doi.org/10.1007/978-3-319-77809-9_3) contains supplementary material, which is available to authorized users.

an insulation material may be well known and accurately modeled, how can one model the actions of building occupant with certainty? Does an occupant show up within a Space for the same period of time each day, and how many are there? Can we predict the level of physical activity an occupant undertakes? How will the occupant operate lights and other equipment? These are only a few factors that drive uncertainty in an analytical process that requires unambiguous numerical input of occupant behavior for 8760 h of each simulated year.

Before you give up and throw this book in the rubbish bin, consider a few key points:

1. As mentioned in Chap. 1, comparative analysis is one of the most important capabilities offered by building energy modeling. As long as occupant behavior is held constant across simulations, uncertainty in occupant behavior will lead to systematic error, which in general, does not invalidate comparisons between multiple energy simulations. The energy impact of changes in occupant behavior can also be modelled as long as the assumptions behind changing the simulation input are well understood.
2. Guidelines like ASHRAE Standard 90.1, ASHRAE Standard 189.1, and California's Title 24 compliance modeling approach are prescriptive about many inputs for energy modeling including Space load assumptions. These guidelines have been informed by surveys like RECS (Residential Energy Consumption Survey), CBECS (Commercial Building Energy Consumption Survey), and CEUS (California End Use Survey) that provide some insight into how different building types are most frequently used. Use of prescriptive input sets is certainly no guarantee that modeled and as-operated energy performance will agree for a specific building, but it does improve consistency in the modeling process.
3. Unlike the previous generation of BEM tools, parametric analysis of multiple OpenStudio simulations (discussed in Chap. 7) enables us to perform Monte Carlo sampling over uncertain inputs. Adopting statistical methods allows the analyst to evaluate distributions of expected performance, which frankly is a more responsible approach than reporting a single, expected savings number to a stakeholder.
4. For existing buildings, calibration (tuning) of building models against measured consumption data is a recommended option to reduce uncertainty in key inputs. OpenStudio makes the calibration process for existing buildings relatively easy.

This Chapter focuses on modeling with standardized inputs (item 2 above) in order to create more consistent modeling outcomes. The approaches suggested in bullet points 3 and 4 will be key topics that we shall revisit in Chap. 7.

## 3.2 Space Types

Just as Construction Sets are comprised of Constructions, which are in turn comprised of Materials, OpenStudio defines Space Types in terms of smaller building blocks that may be assembled flexibly to describe a variety of programmatic activities. In general, OpenStudio Space Types are defined by a set of thermal loads and

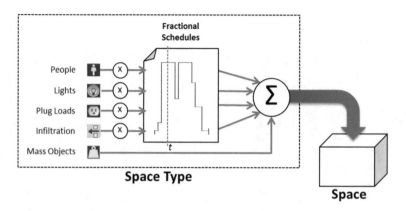

**Fig. 3.1** OpenStudio Space Type conceptually applied to a Space

schedules. In aggregate, these loads consume energy and add heat to a Space as a function of time as shown in Fig. 3.1.

Space Types make it easy to define representative loads and schedules once, in order to apply them quickly and consistently across Spaces with similar activities. A Space Type may apply to a single Space within a building, or it may be applied to multiple Spaces if similar activities take place within them. As with Construction Sets, Space Types may also be imported from other OSMs to speed up the modeling process even further.

### 3.2.1   Schedules

As we have established, individual loads within Spaces are most often a strong function of occupancy, which usually varies with the time of day and day of the week. In terms of modeling approach, this means that we will need to capture occupant, lighting, and equipment schedules if we are to reasonably describe energy uses and their associated thermal loads. The Schedules (▣) Tab in the OpenStudio Application shown in Fig. 3.2 is used for this purpose. New Schedule Sets may be created with the ▣ Button, the ▣ Button is used to duplicate Sets for editing, and the ▣ Button deletes Sets. The ▣ Button "purges" unused Schedule Sets from your Model and is useful for tidying up occasionally.

The Schedules (▣) Tab consists of two Sub-Tabs. The ⸢Schedule Sets⸥ Sub-Tab is used to define collections of Schedules that apply generally to a Space Type. The illustrated example contains Schedule Sets for office break and conference rooms, lobby and open office space. Each Schedule Set may include any or all of the ten categories shown. Individual schedules within a Set may be distinct or duplicative. For example, Occupancy, Lighting, and Electric Equipment Schedules might be the same in an office with staff who diligently turn off lights and equipment whenever they leave a Space. However, different schedules might be used to indicate that occupants leave some lights and equipment on when they leave for lunch.

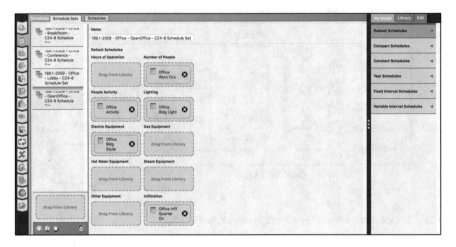

**Fig. 3.2**  Schedule Set for an Office Break Room

The right-hand pane lists the six types of Schedules that OpenStudio supports. Compact, Constant and Year Schedules are identical to the EnergyPlus objects documented online.[1] Fixed and Variable Interval Schedules are used to capture schedules from actual recorded data, these are similar to the Schedule:File Object in EnergyPlus but the data is contained in the OSM rather than an external Comma Separated Value (CSV) file. The final OpenStudio Schedule is the Ruleset Schedule. This Object is unique to OpenStudio and is automatically converted into a Year Schedule for use by EnergyPlus prior to simulation.

Ruleset Schedules are created and edited in the ▭ Sub-Tab as shown in Figs. 3.3 and 3.4. These particular schedules are "fractional," meaning they vary from 0 to 1 and are used as multipliers on the maximum expected occupancy in a Space to create a time varying number of people. Similar Schedules are used to modulate the total power consumed by lighting, electric Equipment, etc. The Schedule Editor enables visual editing of a variety of schedules and will be discussed in Sect. 3.4.

### 3.2.2  Load Definitions

Space Load definitions are entered using the Loads (▣) Tab, and fall into a number of categories including lighting, miscellaneous electric,[2] gas, steam, and other fuel Equipment, people, and hot water uses. Space load definitions describe a particular type of load. We will use Space load instances in Sects. 3.2.3 and 3.3 to actually assign loads of these types to Space Types and Spaces. Lighting loads may include

---

[1] http://bigladdersoftware.com/epx/docs/8-7/input-output-reference/group-schedules. html#group-schedules.

[2] Pieces of Miscellaneous Electric Equipment are sometimes referred to as "MELs" or Plug Loads.

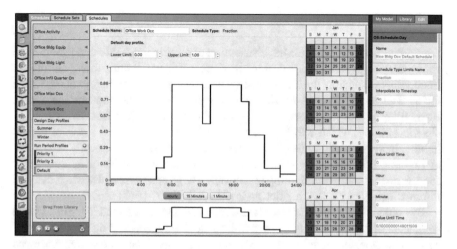

**Fig. 3.3** An office Ruleset Schedule defined for workdays

**Fig. 3.4** An office Ruleset Schedule defined for Saturdays, and Sundays

individual table or desk lamps, arrays of linear fluorescent tubes, emergency exit lights, high intensity discharge (HID) lamps in high bay installations, and many more. Copy machines, coffee pots, microwave ovens, simple refrigerators, laptops, televisions, video game consoles, hair dryers, etc. are all examples of electric plug loads. Common gas loads include ovens or cook tops. In OpenStudio, occupants, infiltration, and internal mass objects are grouped along with these loads but are special cases and will be discussed in subsequent sections. Like most of the Application Tabs, use the ⬚ , ⬚ , ⬚ , and ⬚ Buttons to create, duplicate, delete, and purge Loads.

Depending upon use, available data, etc. the power consumed by most loads is typically entered in OpenStudio in one of three ways:

- Rated power consumed by an individual unit (e.g. single laptop or television) within a Space,
- Rated power consumed per unit of floor area in a Space, or
- Rated power consumed per occupant within a Space.

### 3.2.2.1  Lights and Luminaires

EnergyPlus does not simulate the distribution of electric lights, nor can it be used to verify that a given lighting design provides sufficient illumination. During an EnergyPlus simulation, the electric energy used by lights and luminaires is accounted for as well as the thermal impact of this energy use on the surrounding Space. EnergyPlus does account for the use of daylighting controls to offset energy used by electric lighting, which will be discussed in Chap. 8. Figure 3.5 shows a typical lighting load entered in units of watts per square foot. This is frequently referred to as Lighting Power Density (LPD) and is a common way of characterizing loads such as linear fluorescent lighting that covers a large area; especially when detailed information about the lighting design is not available.

By comparison, individual desk lamps might be best represented in terms of watts per person. If a detailed audit has determined the exact number of lighting fixtures of a particular type, then lighting may be specified in terms of the rated power consumed by an individual unit. In all cases, a multiplier is applied to the space load instance to describe the number of units present in the Space Type or Space. Regardless of how the load is characterized, individual loads are subsequently multiplied by an associated

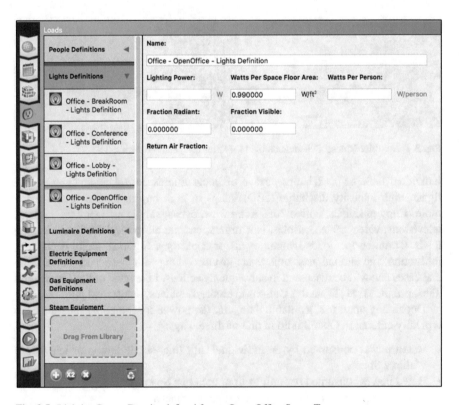

**Fig. 3.5** Lighting Power Density defined for an Open Office Space Type

fractional schedule as described in the previous section. This enables us to simulate the time varying nature of power consumption with the added flexibility of separating the magnitude of loads from how frequently they are used. For lighting, all energy used by the fixture is rejected to the environment (there is no useful mechanical work done and very little energy storage). The mechanisms for heat transfer from Lights and Luminaires are:

- **Fraction Radiant** – portion of energy radiated to the Space's Surfaces as long-wave radiation,
- **Fraction Visible** – portion of energy radiated to the Space's Surfaces as short-wave radiation,
- **Return Air Fraction** – portion of energy rejected to air leaving the space to an air handler, and
- **Fraction Convected** – portion of energy rejected to the Space's air volume.

The fraction convected is not entered directly, rather it is calculated from:

$$f_{convected} = 1 - f_{radiant} - f_{visible} - f_{return\,air}$$

The EnergyPlus Input Output Reference has typical values for these fractional energy factors for different lighting categories under the section for Lights.

Luminaires do not exist as EnergyPlus Objects, and they are a unique to OpenStudio's Object Model. Luminaires are meant to represent individual lighting fixtures and are specified in terms of lighting power per unit. Luminaires also have the ability to be positioned within a Space. This is meant to support simulation of the distribution of electric lighting via Radiance in future versions of OpenStudio.

Like Materials, Constructions, and Construction Sets; Light and Luminaire definitions may also be imported to the current Model from an external Library. Load definitions obtained via Library import contain pre-populated power and heat rejection values, however the modeler should always consider the appropriateness of these values for the specific application. The reader is referred to sources such as the ASHRAE Fundamentals Handbook[3] or The Lighting Handbook published by the Illuminating Engineering Society (IES), as well as manufacturer data sheets for input values DiLaura et al. (2011).

### 3.2.2.2   Electric, Gas, Steam, and Other Equipment Loads

Similar to lighting, EnergyPlus does not represent individual electric, gas, steam, or other Equipment within Spaces. Most people would be concerned if their favorite video game console was replaced by a toaster. However, to EnergyPlus, these are simply devices that consume electricity and reject heat into the Space. The only way that EnergyPlus differentiates between a blender and an electric kettle is by the maximum power draw, schedule of operation, and the mechanisms of heat transfer

---

[3] ASHRAE (2013).

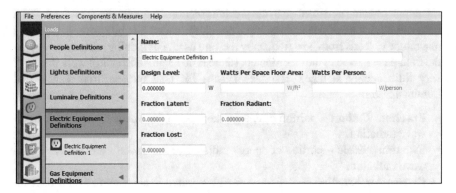

**Fig. 3.6** Example of an Electric Equipment Load

into the Space. Like Lights, Equipment may be quantified by power per unit, floor area, or person (Fig. 3.6). However, unlike Lights, Equipment may consume different types of fuel (e.g. electricity, gas, steam, or other). Each fuel type is accounted for separately in the simulation results.

Note that EnergyPlus allows for a Hot Water Equipment Object. This Object is included in the OpenStudio SDK but has been deprecated from the OpenStudio Application in preference to the more flexible Water Use Equipment Object discussed later. Gas Equipment allows the user to specify a carbon dioxide generation rate for air contaminant studies, this is an advanced topic beyond the scope of this text. Other Equipment allows the user to select from less common fuel types – e.g. propane, fuel oil, etc. The mechanisms of heat transfer allowed for Equipment objects are:

- **Fraction Latent** – portion of energy added to the Space's air volume as moisture;
- **Fraction Radiant** – portion of energy radiated to the Space's Surfaces as long-wave radiation;
- **Fraction Lost** – portion of energy that does not impact the Space's heat balance, by performing useful mechanical work or via rejection outside of the Space; and
- **Fraction Convected** – portion of energy rejected to the Space's air volume.

Fraction Convected is not entered directly and is calculated according to:

$$f_{convected} = 1 - f_{latent} - f_{radiant} - f_{lost}$$

### 3.2.2.3  People

People represent very significant thermal loads in Spaces but are treated a bit differently than lights or Equipment. Like other Loads, occupancy may be characterized by the number of people present within a Space, or in terms of occupant density as

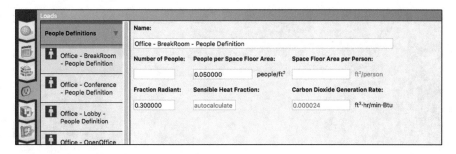

**Fig. 3.7** People definition for an Office Break Room Space Type

shown in Fig. 3.7. People are also described by a fraction of radiant energy they contribute to a space. The remainder of the heat they reject is split between sensible and latent heat addition. EnergyPlus is able to automatically calculate the split between sensible and latent heat gain from occupants based on typical metabolic rates given the level of heat rejection. Astute readers may note that there is no field to enter the "rated" power of a human in this window, so how does EnergyPlus know how much heat each person is rejecting to use along with the fractional occupancy Schedules?

Human power consumption is represented by an additional activity Schedule to better reflect the time varying nature of human activity. Figure 3.8 illustrates a typical activity Schedule for a medium office in the Schedules (▣) Tab. In practice, activity Schedules are often entered as constant values, as shown in this example, to reflect that the type of activity in the space is constant while the number of occupants vary according to the occupancy schedule. However, the software allows for the Modeler to capture a half-hour of calisthenics or naptime, depending upon the culture prevalent in a building, independent of the occupancy schedule. Table 3.1 contains typical power consumption estimates for a range of adult activity.

Additional activity values may be found in the Energy Plus Input Output guide.[4]

### 3.2.2.4   Water Use Equipment

Water Use Equipment represents uses of water within a Space used for showers, cooking, washing, etc. Water Use Equipment is defined in terms of a peak flow rate, which is modulated by a fractional schedule. A Temperature Schedule dictates the temperature of the water. The data entry form for a piece of Water Use Equipment is shown in Fig. 3.9.

The method of accounting for water heating energy depends upon whether or not heating Equipment has been installed elsewhere in the building Model, as will be discussed in Chaps. 4 and 5. If heating Equipment has been specified, then the target

---

[4] http://bigladdersoftware.com/epx/docs/8-7/input-output-reference/group-internal-gains-people-lights-other.html#field-activity-level-schedule-name.

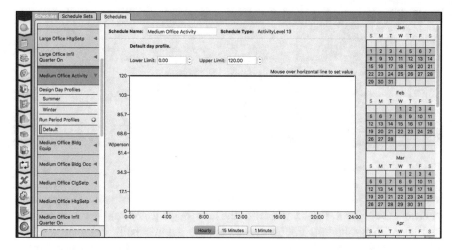

**Fig. 3.8**  Occupant activity Schedule for an office

**Table 3.1**  Typical power consumption associated with various activities (ASHRAE 2013, Table 1, p. 18.4.)

| Activity | Average Power Consumption (W) |
|---|---|
| Sitting | 97 |
| Moderate office work | 130 |
| Light factory work | 220 |
| Heavy factory work | 425 |
| Exercise | 586 |

temperature is achieved by mixing hot and cold water at the fixture. If the Water Use Equipment has not been connected to a heating system in the simulation, then it is simply provided at the requested temperature and the energy required to heat mains water to the given temperature is not tracked.

In either case, the water consumption is accounted for in the simulation along with any heat rejected to the Space. The mechanisms for heat transfer to the Space are sensible heat addition as well as latent moisture addition, any remaining heat is assumed to be convected out of the space down the drain. For maximum flexibility, the fraction of energy rejected to the Space in sensible and latent form are specified by a schedule rather than a fixed value. Water use Equipment will be discussed further in Chap. 5.

### 3.2.2.5  Infiltration

Infiltration is unconditioned outdoor air that enters a space from the outside through gaps in the envelope, doors or windows that are periodically opened. Design Specification Outdoor Air is unconditioned, fresh outdoor air intentionally brought into the building for occupant health. Because infiltration and design specification

**Fig. 3.9** Example water use equipment definition

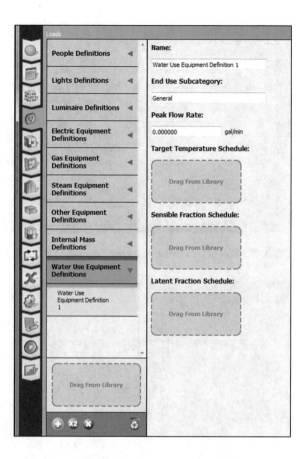

outdoor air increase or decrease heat and humidity in a Space, they are considered a thermal Load that must be actively managed by HVAC systems. Unlike other Loads, infiltration is defined directly in the Space Types (▣) Tab as shown in Fig. 3.10.

Note that in each of the four Space Types shown in the above example that three types of infiltration Objects may be included:

- Design Specification Outdoor Air,
- Space Infiltration Design Flow Rates, and
- Space Infiltration Effective Leakage Area.

The first is associated with ASHRAE Standards 62.1 and 62.2, which prescribes minimum outdoor ventilation rates for occupant health and comfort and will be discussed further in Chaps. 4 and 5. The latter two Objects represent different models for <u>unintended</u> air infiltration. Modelers will most often select only one of these last two Objects to represent Space infiltration. Figure 3.11 shows some of the input arguments for the Design Flow Rate method of calculation.

Figure 3.12 illustrates selection of the Effective Leakage Area Object from the Library after dragging and dropping it onto a Space Type. Note that dragging infiltration

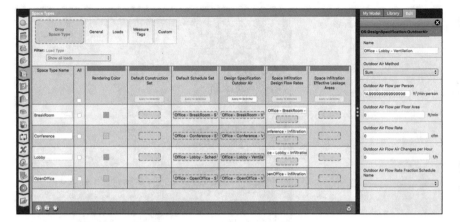

**Fig. 3.10** Outdoor air ventilation for an Office Lobby Space Type

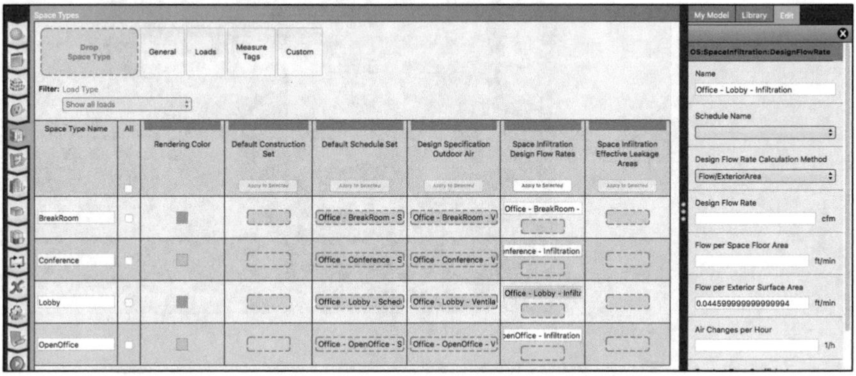

**Fig. 3.11** Infiltration Flow Rate for an Office Lobby Space Type

objects in from the library is the only way to create infiltration and design application outdoor air objects in the OpenStudio Application. The reader is directed to the EnergyPlus Input Output Reference Guide for more information on these Objects[5] and appropriate selection of input parameters.

### 3.2.3   Building a Space Type

The previous three Figures were our first introduction to OpenStudio's Space Type (🔲) Tab. The "General" view allows the user to drag Space Types in from a Library, as well as to create, duplicate, delete, or purge Space Types with the appropriate

---

[5] http://bigladdersoftware.com/epx/docs/8-0/input-output-reference/page-018.html#group-airflow.

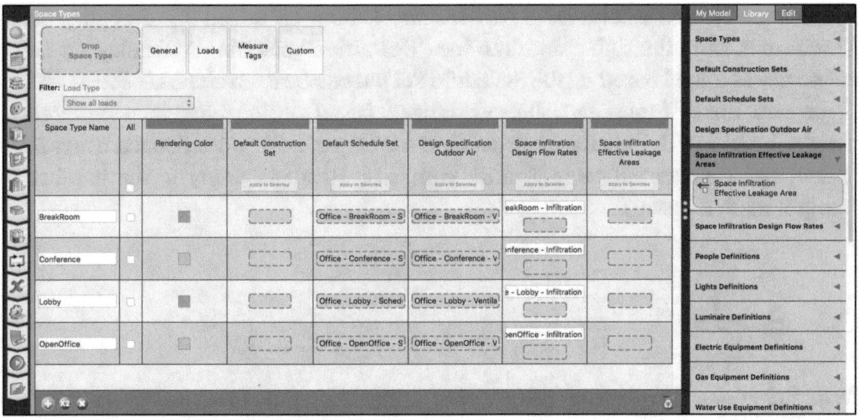

**Fig. 3.12** Adding an Infiltration Effective Leakage Area Object

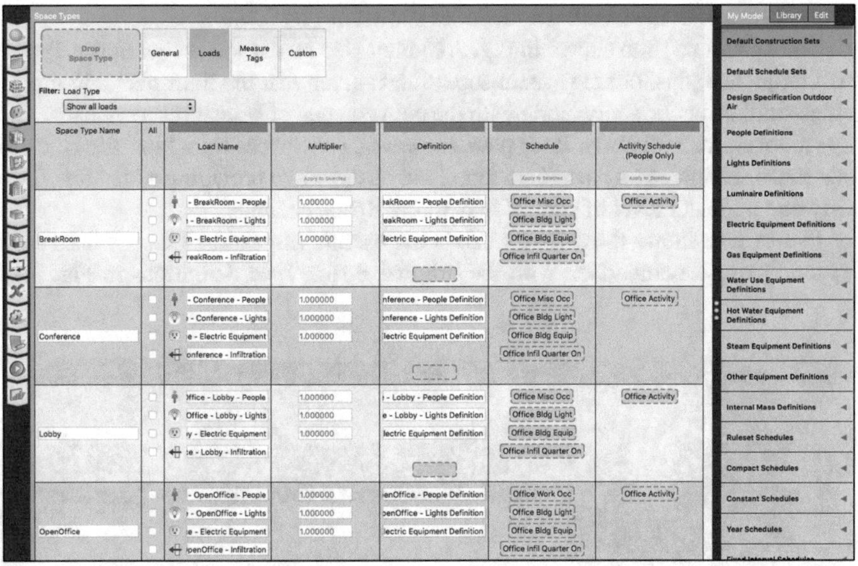

**Fig. 3.13** Adding loads to define four office Space Types

Button. This first view allows the user to assign custom colors, Construction Sets, Schedule Sets, and infiltration Objects that will be used whenever the Space Type is assigned to a Space. Note that in Fig. 3.12 no Construction Set has been assigned to any of the Space Types. Recall from the previous Chapter that in the absence of a specific Construction Set definition, OpenStudio will look to the Object's parent (in this case the Building Story) for guidance.

Clicking on the Loads Button at the top of the window switches to a breakdown of Loads within each Space Type as shown in Fig. 3.13. Infiltration will show up automatically, and additional Load Instances may be added by dragging Load Definitions from the right hand into the "Definition" column. Schedules are automatically assigned based on the Schedule Set but can be overridden if needed. Space Type and Load Names may be customized based on the modeler's preference. Lastly, Load Multipliers may be specified for Objects that were defined using individual rated power values (e.g. a multiplier of 10 might apply to the number of computers in a Space).

## 3.3   Spaces

With Schedules, Loads, and Space Types defined, it's time to assign Space Types to specific activity areas within the building. In the previous Chapter, we used the floor plan editor in the Geometry (▣) Tab to create Spaces. As we with Thermal Zones in Sect. 2.8.7, the Assignments Sub-Tab of the floor plan editor may be used to assign Space Types to Spaces within a model as shown in Fig. 3.14.

The Spaces (▣) Tab shown in Fig. 3.15 may also be used to assign Space Types. At first glance, this Tab may seem superfluous given that the floor plan editor provides similar functionality along with the convenience of being able to visualize the Space locations within the floor plan. However, the Spaces Sub-Tabs allows us to edit multiple Spaces on each Floor simultaneously, while providing added functionality and a greater level of control in defining Space contents.

Figure 3.16 shows the ⌈Loads⌉ Sub-Tab being used to inspect the Loads within each Space. A quick comparison with the original Space Type definitions in Fig. 3.13

**Fig. 3.14** Assigning Space Types using the Geometry editor

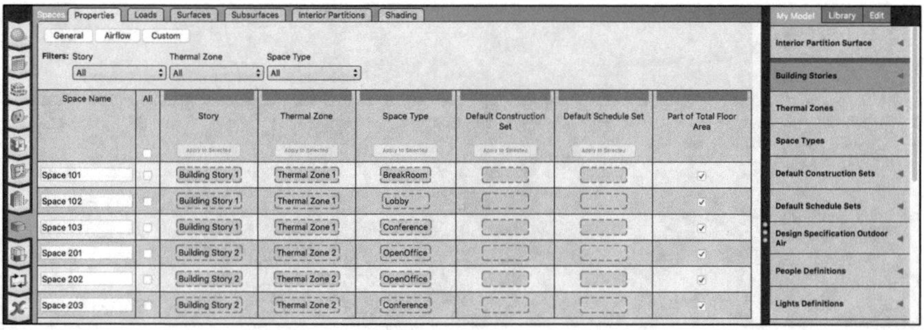

**Fig. 3.15** Space assignment Sub-Tab

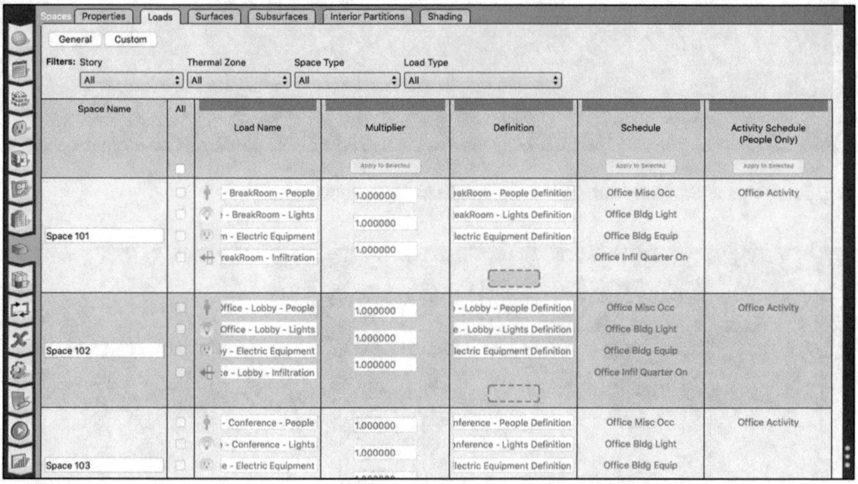

**Fig. 3.16** Space load assignments defaulted from Space Type definition

confirms that the loads prescribed by each Space Type were applied correctly. One subtle distinction between these two views is the text color. The text in the Space Type definitions is black, whereas the text in the actual Spaces is green. This is OpenStudio's visual cue that data is being inherited and is used throughout the Application. This particular Sub-Tab allows the user to drag in additional loads that will show up in black text, which may be unique to an individual Space that is otherwise well described by a Space Type. By the same token, checkboxes next to inherited rows may be used to select individual Loads for deletion with the ⊟ Button. Like Construction Sets, Space Types allow the user to quickly assign activities and loads from manually assembled definitions or Libraries, but they are not restrictive, and Spaces may be customized when necessary.

The Surfaces and Subsurfaces Sub-Tabs shown in Figs. 3.17 and 3.18 also illustrate the data inheritance concept with Construction Sets. In this case, appropriate Constructions have been applied to each Surface and Sub-Surface based on the

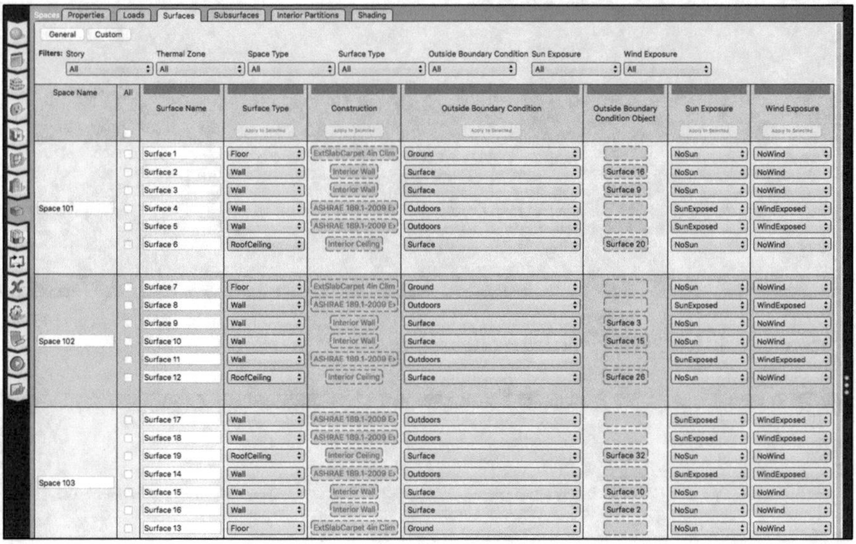

**Fig. 3.17**  Space Surfaces Sub-Tab with defaulted Construction Set

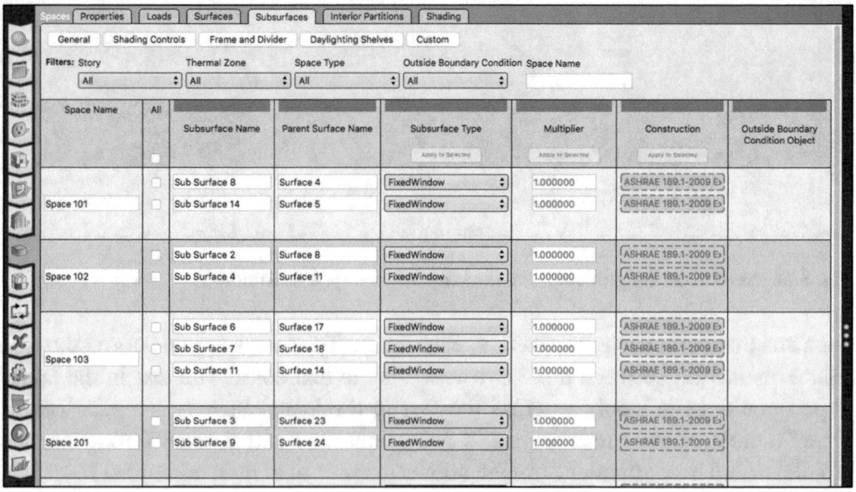

**Fig. 3.18**  Space Sub-Surfaces Sub-Tab

Surface Type. Unique Constructions may be dragged in to replace inherited Constructions when required.

One type of Load that was briefly mentioned earlier in the Chapter is an "Internal Mass Object." These Objects may be added to Space Types like other Loads as shown in Fig. 3.19. Internal Mass Objects don't consume energy or radiate heat in the same sense as other Objects like people or Equipment, nor do they have associated Schedules. Instead, the Objects add thermal capacitance to the spaces they

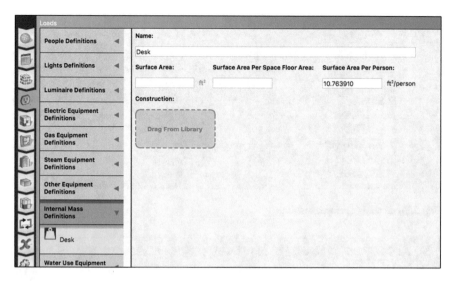

**Fig. 3.19**  Defining an Internal Mass Object for a Space

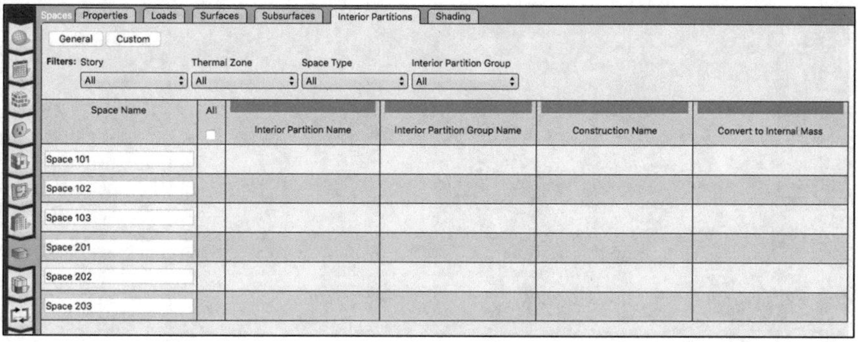

**Fig. 3.20**  Interior partitions Sub-Tab

occupy; that is, they store and release heat, adding to the dynamic behavior of the Space. The Object's Construction and surface area dictate this behavior. Internal Mass Objects can provide a very significant effect in Spaces used for storage (e.g. warehouses with heavily laden shelves).

We mention Internal Mass Objects here because they are closely connected with the [Interior Partitions] Sub-Tab shown in Fig. 3.20. Interior Partitions are Surfaces that exist within a Space to represent furniture, or other objects. Interior Partitions are included in detailed renderings using Radiance but do not define boundaries between Spaces or with the exterior environment. However, they may act as Internal Mass Objects, storing and releasing heat within a Space. Interior Partitions will generally be specified along with the Space geometry with an associated Construction. Checking the "Convert to Internal Mass" column tells OpenStudio to include it as part of the overall thermal mass within the Space.

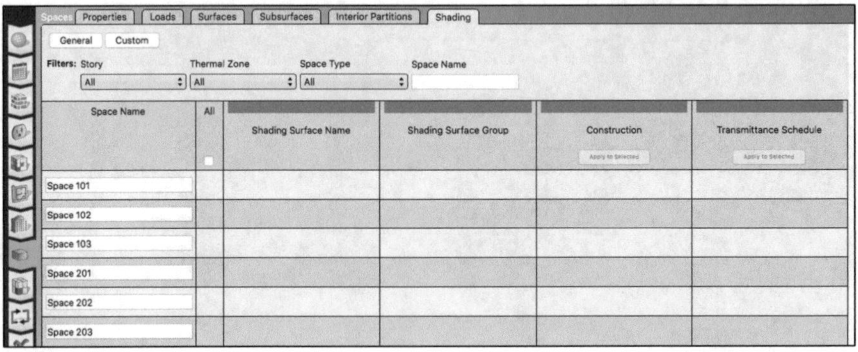

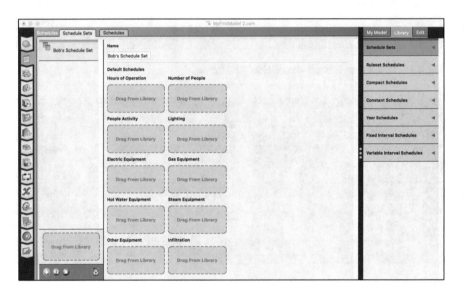

**Fig. 3.21**  Shading Surfaces Sub-Tab

**Fig. 3.22**  Creating a new Schedule Set

The last Sub-Tab associated with Spaces is ▭Shading, shown in Fig. 3.21. The significance of Shading Surfaces as it relates to solar radiation exposure and daylight availability will be discussed in Chap. 8.

## 3.4  Checkpoint Three: Ideal Air Load Simulation of a Small Model

Now it's time to revisit the first Model we built in Checkpoint One. Recall when we last worked with that Model; it was a simple box comprised of a single Space and Thermal Zone with no Space Type definition and no HVAC system. The Space temperature was strictly a function of ambient heat transfer through the envelope. In this exercise, we will define a Space Type around a single unfortunate occupant named "Bob" who is forced to use the small, unconditioned building as his office. If we're feeling generous, perhaps we'll try to provide some rudimentary air conditioning for him as well. Open up the last saved version of that Model, and let's get to work.

### 3.4.1   Creating a Schedule Set

Open the Model and create a new Schedule Set named "Bob's Schedule Set" with the ⊡ Button in the Schedules (▣) Tab as shown in Fig. 3.22. Leave it blank for the moment and proceed to the ⌗Schedules Sub-Tab.

Use the ⊡ Button to create new Ruleset Schedules. A dialog pops up allowing the user to select the type of Schedule as shown in Fig. 3.23. Create the schedules listed in Table 3.2.

Now we need to edit the Schedules themselves. Select one of the Schedule Objects and note that each allows the user to specify Summer and Winter design day profiles and a Default Run Period Profile. Ignore the design day profile options and select the Default Run Period Profile that has a light blue band next to it. This activates a graphical interface for editing our Schedules as shown in Fig. 3.24. Recall from Sect. 3.2.2.3 that Activity Schedules like Bob's are often set as constant values throughout the year, relying on a fractional occupancy Schedule to modulate them. If you would like to make Bob lazier, click on the constant line and drag it down. You may also hover the mouse over the bar, type in a number and press enter to set a specific value. Let's leave Bob's Activity level at 100 W for the purpose of this exercise.

Having set Bob's typical energy consumption, we now need to specify a working schedule for his thermal torture chamber – aka office. Select Bob's Schedule to produce the Default Schedule shown in Fig. 3.25. The "tip list" and Fig. 3.26 will help you understand how the Schedule editor works.

**Useful Schedule Editor Tips:**
- Double clicking on horizontal lines splits them into multiple segments.
- Horizontal segments may be dragged up and down or set to a specific value by typing a number and pressing enter.
- Double clicking on vertical lines deletes segments.
- Vertical segments may be dragged back and forth.
- The Hourly, 15 min, and 1 min selectors at the bottom of the window allow for finer resolution along the time axis.

Notice that the light blue color band next to Default matches the color of each day on the calendar to the right of the window. This indicates that the Default schedule will be used 365 days of the year. We can allow Bob to take days off by specifying

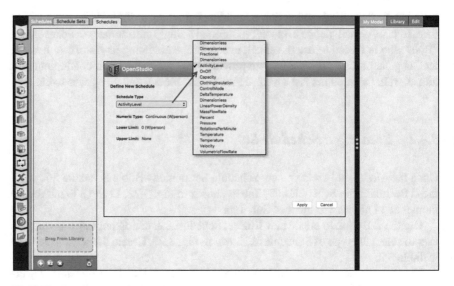

**Fig. 3.23** Creating a new Ruleset Schedule

**Table 3.2** Schedule Types and names for Checkpoint 3 exercise

| Schedule Type | Name |
| --- | --- |
| Activity | Bob's activity |
| Fractional | Bob's schedule |
| Temperature | Heating thermostat |
| Temperature | Cooling thermostat |

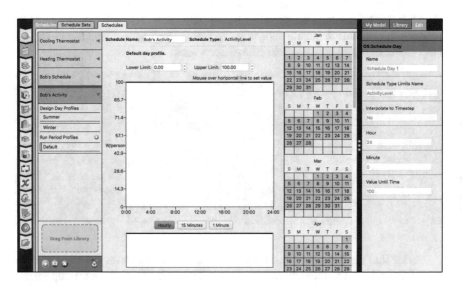

**Fig. 3.24** Editing Bob's activity Schedule

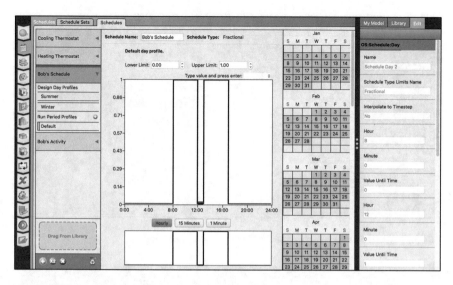

**Fig. 3.25** Bob's default Occupancy Schedule

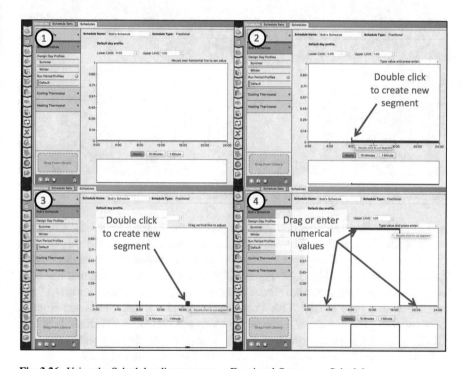

**Fig. 3.26** Using the Schedule editor to create a Fractional Occupancy Schedule

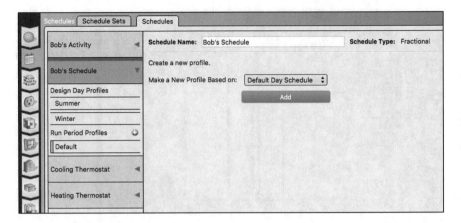

**Fig. 3.27** Creating a new Run Period Profile in a Ruleset Schedule

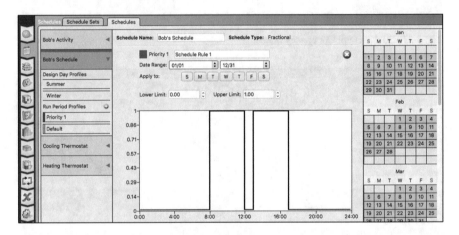

**Fig. 3.28** A new profile for Bob's Schedule copied from the default Profile

additional Profiles and selection rules as part of the Ruleset Schedule. Click the small ▣ next to "Run Period Profiles" to add an additional Profile. We are given the option to create an entirely new Profile or copy from another Profile in the Ruleset Schedule as shown in Fig. 3.27. Figure 3.28 shows a second Profile that has been copied from the default Profile. Notice that our new Profile is called "Priority 1" and marked with a purple band. This Profile also includes a date range and Buttons corresponding to the days of the week that was not present in the Default Profile. Clicking on the Buttons corresponding to Saturday and Sunday and setting the schedule to a uniform 0 fraction leads to the Profile shown in Fig. 3.29.

Additional Profiles may be added with their own date ranges and days of the week to reflect holidays, cleaning crew occupancy, and more. Each new Profile is assigned its own priority number. The highest priority (lowest number) always wins

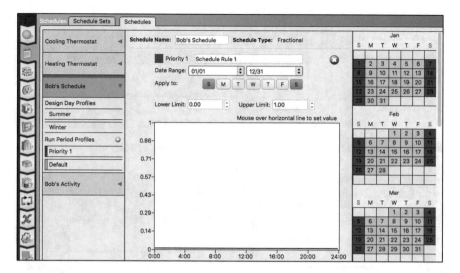

**Fig. 3.29** A higher priority Ruleset Schedule profile for weekends

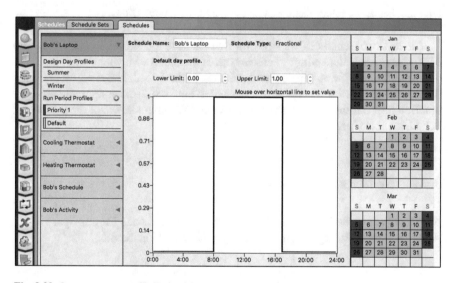

**Fig. 3.30** Laptop usage profile derived from Occupancy Schedule

out – another example of data inheritance in OpenStudio. Checking color codes on the calendar is always recommended to ensure that a Ruleset Schedule will be applied as expected.

Now use the ▣ Button to duplicate Bob's Schedule. Rename it "Bob's Laptop," and edit the Default Profile so it looks like Fig. 3.30. This reflects the likelihood that Bob leaves his Laptop running in the Space when he slips out during his lunch hour.

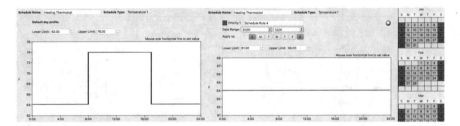

**Fig. 3.31**  Setting a Heating Thermostat Ruleset Schedule for Bob's Office

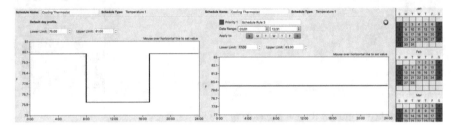

**Fig. 3.32**  Setting a Cooling Thermostat Ruleset Schedule for Bob's Office

The last two Schedules we need for this exercise are shown in Figs. 3.31 and 3.32. Use the Schedule editor to create both of them and save your work.[6]

With all of our Schedules defined, go back to the Schedule Set editor and drag the individual Ruleset Schedules in as shown in Fig. 3.33. This assignment assumes that Bob turns off the lights whenever he leaves. We will make use of the thermostat Schedules elsewhere.

### 3.4.2  Defining Our Loads

Now navigate to the Loads (■) Tab to define all of the Loads for our Space Type. Add Loads for Bob, Lights, and his laptop as shown in Fig. 3.34.

### 3.4.3  Defining and Assigning Our Space Type

With Schedules and Loads defined, it's time to assemble them into a Space Type that we can apply to Spaces within our Building. Use the Space Types (■) Tab to create a new Space Type with the ■ Button. Drag and drop the new Schedule Set into the Default Schedule Set column as shown in Fig. 3.35. Ignore the other

---

[6] The values shown assume that you have set the Application's units preference to English (I-P).

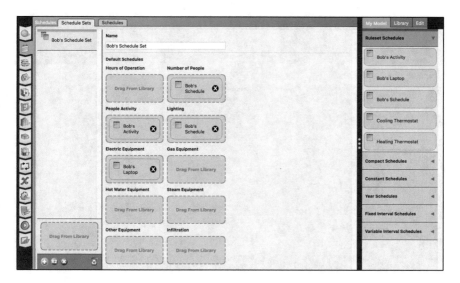

**Fig. 3.33** Defining a Schedule Set for Bob's Office

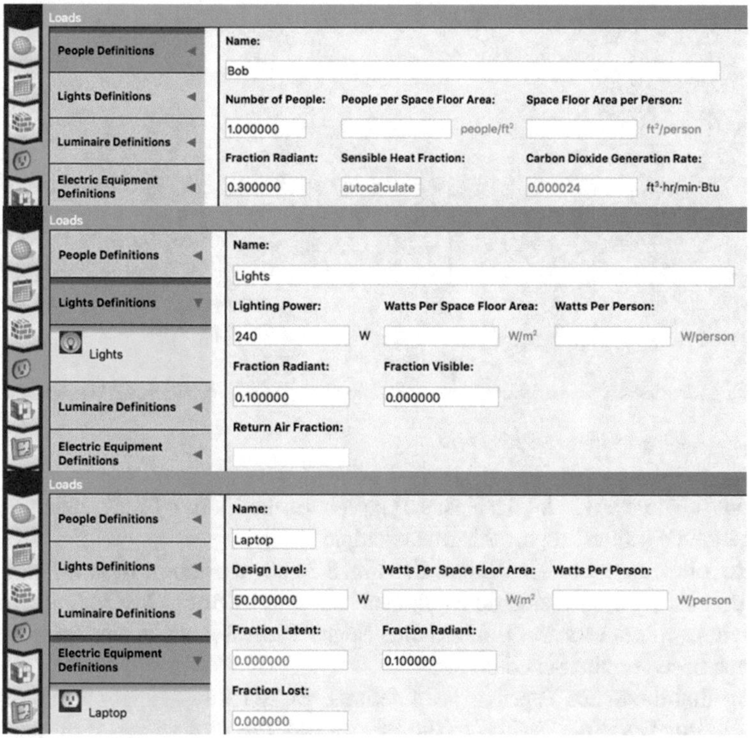

**Fig. 3.34** Defining the Loads for Bob's Office

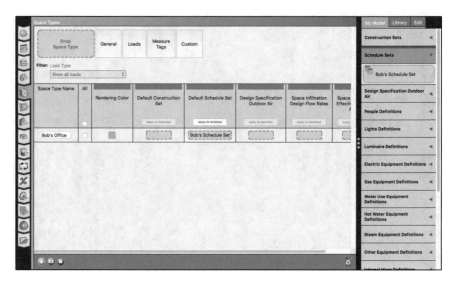

**Fig. 3.35** Creating a new Space Type and Adding a Schedule Set

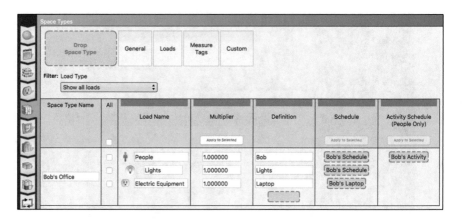

**Fig. 3.36** Adding Loads to a Space Type

columns, since we will let the Space Type inherit the Floor's Construction Set and won't include infiltration for this first example.

Next, click on the Loads Button. Use Fig. 3.36 as an example of how to drag and drop people, light, and electric Equipment definitions from "My Model" into the Definitions column for the Bob's Office Space Type. Schedules are automatically assigned based on the Schedule Set.

Now that the Space Type has been defined, we're ready to apply it to the single Space in our Building. Switch to the Spaces (■) Tab to drag and drop the Bob's Office Space Type onto our Space as shown in Fig. 3.37. Switch to the ⌐Loads¬ Sub-Tab to verify that all of our Loads and Schedules were correctly inherited from the

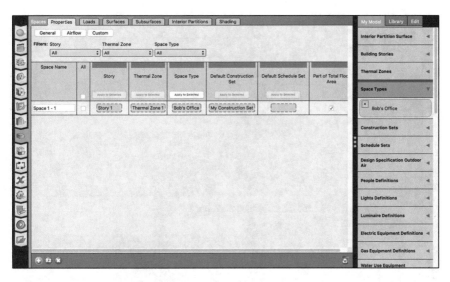

**Fig. 3.37** Assigning our New Space Type to Bob's Space

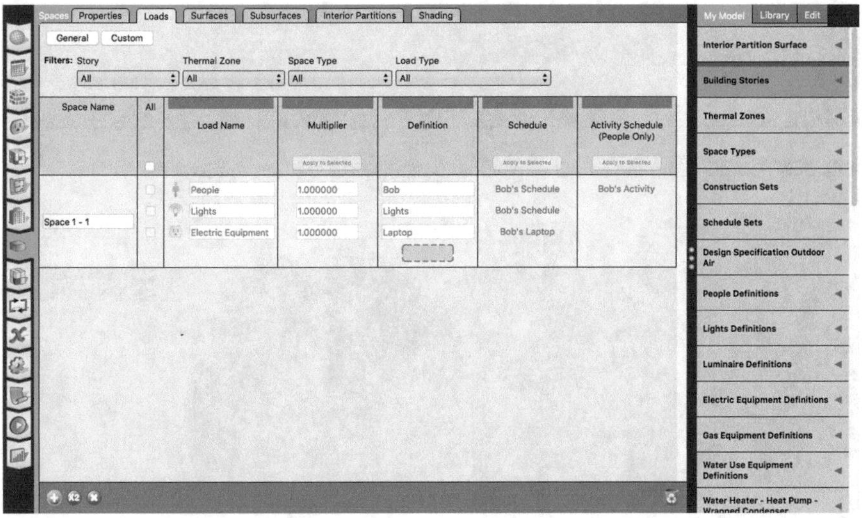

**Fig. 3.38** Inspecting the Loads assigned to Bob's Space

Space Type (Fig. 3.38). If you wish, you may check that Surfaces and Sub-Surfaces were inherited from the Floor/Building as well.

Since this particular building has only a single Space, the extra effort of defining a Space Type isn't necessarily warranted. As mentioned previously in Sect. 3.3, we could also have dragged loads and schedules directly into our Space without relying on data inheritance. The real value of Space Types will become apparent in the next exercise where many Spaces can benefit from common Space Type definitions.

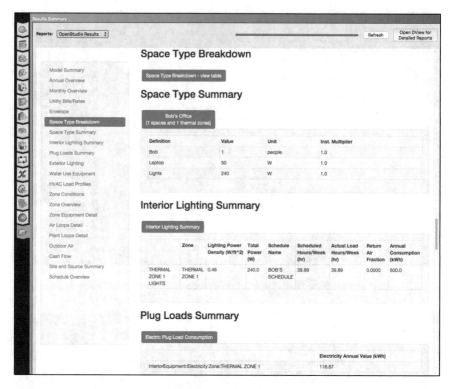

**Fig. 3.39**  Additional content in the OpenStudio report related to our Space

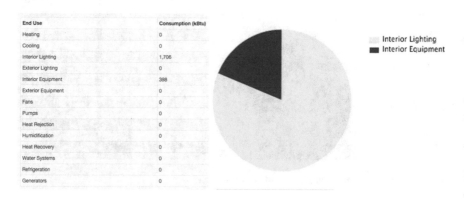

**Fig. 3.40**  Annual end use breakdown for Bob's Unconditioned Office

### 3.4.4 Running the Simulation

We are told that Bob is (very reluctantly) prepared to begin his year-long tenure in his unconditioned office. Let's make sure he shows up by adding "People Occupant Count" as a Timestep variable with the Variables (⬛) Tab. Use the Run (⬛) Tab to run the simulation. When the simulation has completed, switch to the Reports (⬛) Tab to view the OpenStudio standard report. Clicking on the Space Type Breakdown link in the report produces Fig. 3.39 allowing us to quickly verify that all of the Loads we prescribed in "Bob's Office" are present and accounted for.

The Annual Overview link (Fig. 3.40) now includes pie charts showing the end use breakdown for lighting and Bob's Laptop across the entire year. Hovering the

## Monthly Overview

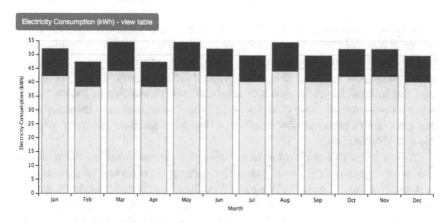

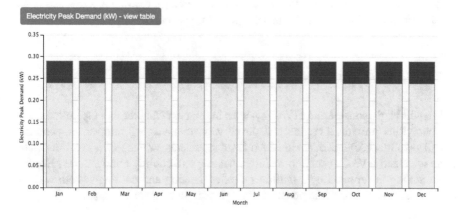

**Fig. 3.41** Monthly end use breakdown for Bob's Unconditioned Office

| Zone | Unmet Htg (hr) | Unmet Htg - Occ (hr) | < 56 (F) | 56-61 (F) | 61-66 (F) | 66-68 (F) | 68-70 (F) | 70-72 (F) | 72-74 (F) | 74-76 (F) | 76-78 (F) | 78-83 (F) | 83-88 (F) | >= 88 (F) | Unmet Clg (hr) | Unmet Clg - Occ (hr) | Mean Temp (F) |
|---|---|---|---|---|---|---|---|---|---|---|---|---|---|---|---|---|---|
| THERMAL ZONE 1 | 0 | 0 | 42 | 169 | 419 | 245 | 342 | 432 | 446 | 420 | 371 | 831 | 708 | 4335 | 0 | 0 | 93.1 (F) |

| Zone | < 30 (%) | 30-35 (%) | 35-40 (%) | 40-45 (%) | 45-50 (%) | 50-55 (%) | 55-60 (%) | 60-65 (%) | 65-70 (%) | 70-75 (%) | 75-80 (%) | >= 80 (%) | Mean Relative Humidity (%) |
|---|---|---|---|---|---|---|---|---|---|---|---|---|---|
| THERMAL ZONE 1 | 1980 | 560 | 486 | 459 | 371 | 359 | 355 | 372 | 356 | 343 | 328 | 2791 | 58.4 (%) |

**Fig. 3.42** Zone Conditions for Bob's Unconditioned Office

mouse over these plots reveals specific numbers, and the blue "view table" Buttons expand to reveal detailed tabular data underlying the charts.

Also of note is the Monthly Overview (Fig. 3.41). These monthly consumption plots are essentially flat, varying only based on the number of workdays that fall in any given month.

Of greatest interest (at least to Bob) is the Zone Conditions section of the report, as it most concisely reflects the cruelty we have inflicted upon him. As evidenced by Fig. 3.42, adding Bob, his laptop, and lighting to the Space has increased the number of hours spent in excess of 88 degrees Fahrenheit by nearly 300 h! On the bright side, the thermal Loads in Bob's office decreased the number of extremely cold hours a small amount. Also of significance is the dramatic shift in humidity within the Space. Bob's presence now results in nearly 3000 additional hours at 80% relative humidity or higher.

Adding insult to injury is the time series plot produced using DView and the simulations' eplusout.sql file. Figure 3.43 compares ambient and interior temperatures over the same time period we plotted in Chap. 2. This new plot shows that Bob did indeed turn up for work each day, and experienced peak temperatures that were nearly ten degrees higher than if he and his equipment weren't in the Space. The authors feel terrible about Bob's deplorable working conditions. Let's see if we can't do something for him in the next part of this exercise.

### 3.4.5   Ideal Air Loads

We will begin considering HVAC systems in the next Chapter, but OpenStudio and EnergyPlus provide a "quick and dirty" way of managing interior temperatures called an Ideal Air Load. An Ideal Air Load represents a Zone that is conditioned by an idealized HVAC system to maintain heating and cooling temperatures. Heating or cooling is supplied by a fictitious district[7] heating and cooling system with no significant sizing or dynamic constraints placed upon its performance.

---

[7] District systems are heating or cooling systems that exist external to the building. Examples include central steam or chiller plants that provide heating and cooling to a campus of buildings.

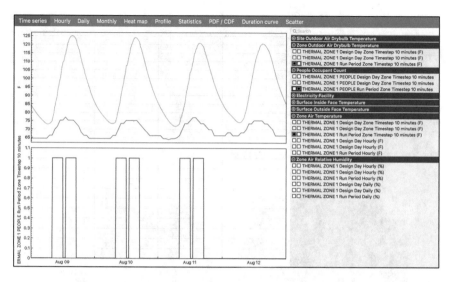

**Fig. 3.43** DView time series plots for Bob's Unconditioned Office

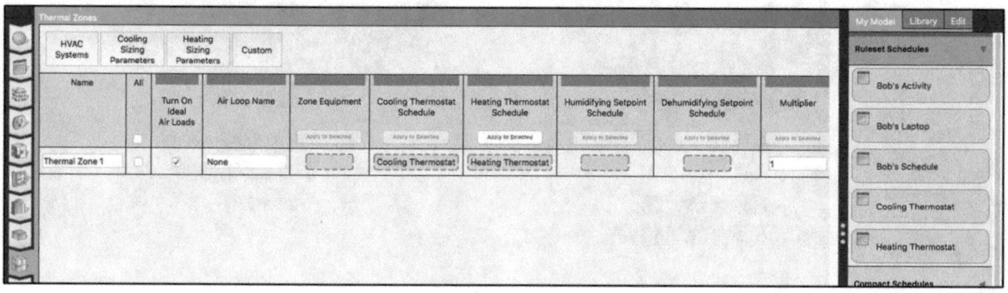

**Fig. 3.44** Adding Thermostat Schedules and Ideal Air Loads to our Thermal Zone

Ideal Air Loads are turned on with a simple checkbox on the Zones (▣) Tab. The Ideal Air Load requires heating and cooling thermostat schedules to function properly. Fortunately, we created them when we built our Space Type definition. Drag the thermostats onto the Zone and check the Ideal Air Loads box as shown in Fig. 3.44. Remember to save your work, and run the updated simulation to see if we've made Bob's life any better.

Jumping straight to the Zone Conditions summary in the OpenStudio report (Fig. 3.45) shows a tremendous change for the better. There are now zero hours in excess of 88 degrees Fahrenheit. The number of chilly hours could be improved by alterations to the heating thermostat schedule, but we don't want Bob to get too complacent. The Annual and Monthly Overview sections of the reports shown in Figs. 3.46 and 3.47 have also changed to reflect the addition of district heating and cooling to condition Bob's Office.

Figure 3.47 represents our closest look yet at end use breakdowns required to heat, cool, and light a building (albeit a simple one) that supports occupant activity. As a building modeler, it is important to inspect these kinds of results to see if they

| Zone | Unmet Htg (hr) | Unmet Htg - Occ (hr) | < 56 (F) | 56-61 (F) | 61-66 (F) | 66-68 (F) | 68-70 (F) | 70-72 (F) | 72-74 (F) | 74-76 (F) | 76-78 (F) | 78-83 (F) | 83-88 (F) | >= 88 (F) | Unmet Clg (hr) | Unmet Clg - Occ (hr) | Mean Temp (F) |
|---|---|---|---|---|---|---|---|---|---|---|---|---|---|---|---|---|---|
| THERMAL ZONE 1 | 0 | 0 | 0 | 0 | 2053 | 665 | 610 | 450 | 420 | 1363 | 1659 | 1540 | 0 | 0 | 0 | 0 | 72.3 (F) |

| Zone | < 30 (%) | 30-35 (%) | 35-40 (%) | 40-45 (%) | 45-50 (%) | 50-55 (%) | 55-60 (%) | 60-65 (%) | 65-70 (%) | 70-75 (%) | 75-80 (%) | >= 80 (%) | Mean Relative Humidity (%) |
|---|---|---|---|---|---|---|---|---|---|---|---|---|---|
| THERMAL ZONE 1 | 0 | 10 | 1770 | 2912 | 766 | 892 | 777 | 1595 | 38 | 0 | 0 | 0 | 47.2 (%) |

**Fig. 3.45** Zone Conditions with Ideal Air Loads

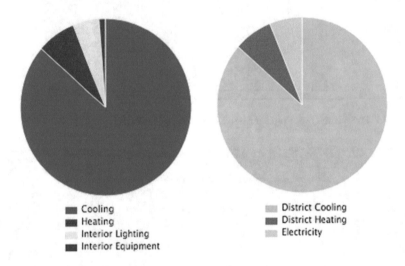

**Fig. 3.46** Annual end use breakdown with Ideal Air Loads

make sense. One feature of these monthly bar charts that should give us pause is the need for simultaneous heating and cooling throughout the year, an indicator of an inefficient design. In the case of an Ideal Air Load system, this can be a result of poorly selected heating and cooling thermostat schedules. We can verify that this is the case by looking at time series plots in DView (Fig. 3.48).

As we add non-idealized HVAC systems to our models, behaviors like simultaneous heating and cooling, unmet hours of operation, and Equipment cycling become more likely. Casting a critical eye at the annual, monthly, and time series data provided by the software is an important role of the energy modeler. Just because the simulation "said so" does not make it correct or optimal. In this case, what would you do to eliminate simultaneous heating and cooling with our Ideal Air Load System? Test your ideas by changing your Model and re-running it.

**Monthly Overview**

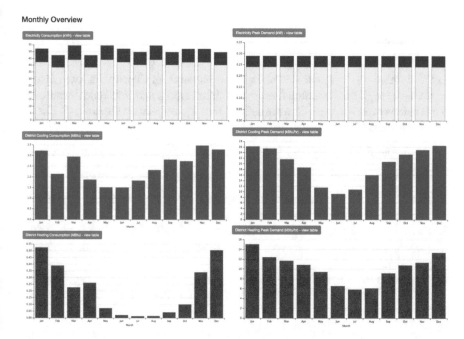

**Fig. 3.47** Monthly end use breakdown with Ideal Air Loads

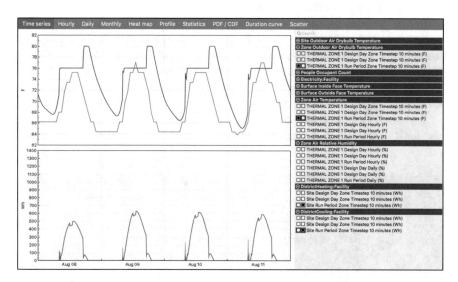

**Fig. 3.48** Time series plots of ambient and interior temperatures alongside district heating/cooling load

## 3.5   Checkpoint Four: Assigning Space Types to a School Model

Let us continue refining our school capstone Model from Checkpoint Two. Instead of going through all the tedious steps of defining Loads and Schedules we used in the previous exercise, we will make use of Libraries to speed up the process. Use the following steps to get started:

1. Open your Primary School Model (or a copy of it) from Checkpoint Two.

   (a) We recommend using "Save As" to preserve your previous work.

2. Pick "Load Library" from the File menu and import *PrimarySchool.osm* to load the Model library with Space Types.
3. Navigate to the Space Type (▣) Tab, select Space Types labeled *90.1–2010 – PriSchl* and drag them into your Model as shown in Fig. 3.49. Rename them if you wish.
4. Navigate to the Building (▪) Tab, select the Classroom Space Type you just added, and drag it to become the Default Building Space Type (Fig. 3.50).
5. Save your Model.

Note that you could have selected any of the Space Types we just added as the default Space Type. In general, you should select the Space Type that is most frequently used in your building – in this case a classroom.

Before proceeding, this is a good opportunity to see how much time we just saved by importing those Space Type Definitions, and also to verify the assumptions we are making with loads, schedules. etc. Use the Loads section of the Space Types (▣) Tab to inspect the definitions for each Space Type as shown in Fig. 3.51. Each definition includes People, Lights, Equipment, Infiltration, and associated Schedules without any significant effort on our part.

We can dig a bit deeper into the underlying definitions using the Schedules (▣) and Loads (◎) Tabs. Take a moment to browse through those sections of the Model

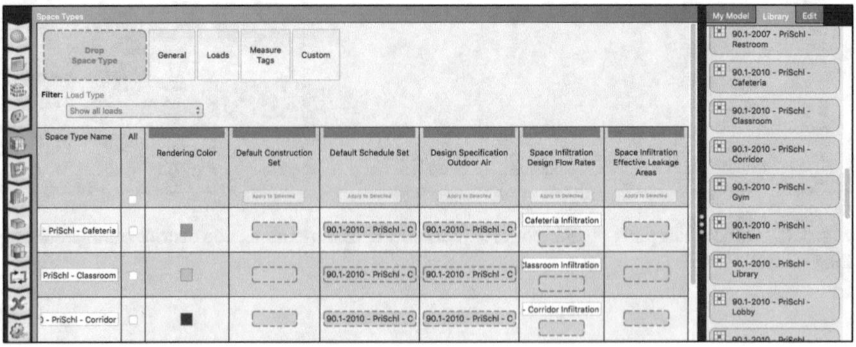

**Fig. 3.49**  Adding Space Types from the Library

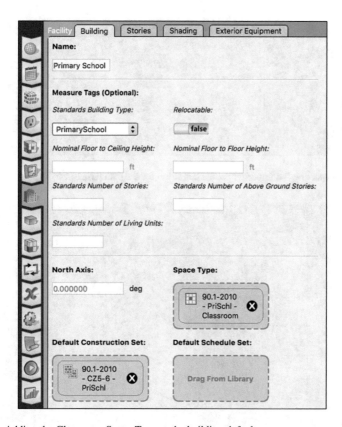

**Fig. 3.50**   Adding the Classroom Space Type as the building default

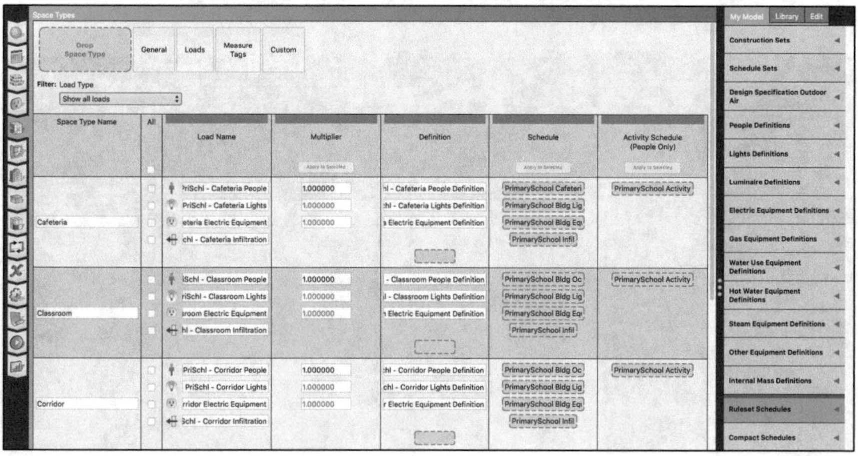

**Fig. 3.51**   Inspecting load definitions for the new Space Types

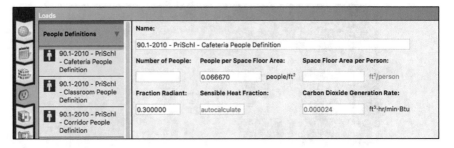

**Fig. 3.52** Inspecting the imported people definition for the Cafeteria Space Type

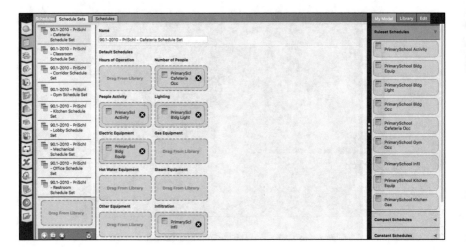

**Fig. 3.53** Inspecting the Imported Schedule Set for the Cafeteria Space Type

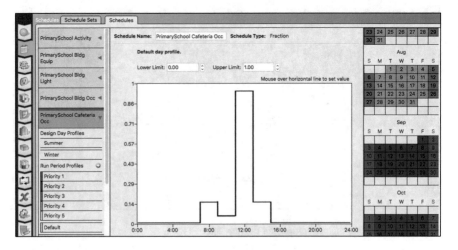

**Fig. 3.54** Inspecting an Imported RuleSet Schedule for the Cafeteria Space Type

to see what data has been added. For example, Fig. 3.52 illustrates the definition of a person in our Cafeteria.

Figure 3.53 highlights the Cafeteria Schedule Set, just one of many we imported. Switch to the [Schedules] Sub-Tab to examine individual Schedules that were added as part of those Sets. For example, Fig. 3.54 illustrates the Cafeteria Occupancy Ruleset Schedule built up from a Default Profile and Five different Priority Profiles representing weekends, the School Year, Holidays, etc. How much time did making use of pre-built schedules save? How many mistakes were potentially avoided?

While we are in this particular Sub-Tab, take a moment to add two additional Ruleset Schedules from the Library you imported. We will make use of: *PrimarySchool HtgSetp* and *PrimaryScool ClgSetp* (Fig. 3.55). As you might guess, these are thermostat schedules that we will use when adding Idea Air Loads to our Thermal Zones.

Having verified the specific Loads and Schedules associated with our new Space Types, it's time to actually apply them to the Spaces in our building. Navigate to the Spaces (■) Tab. Note in Fig. 3.56 that the Classroom Space Type has been automatically assigned to all Spaces in the building per our default assignment. In general,

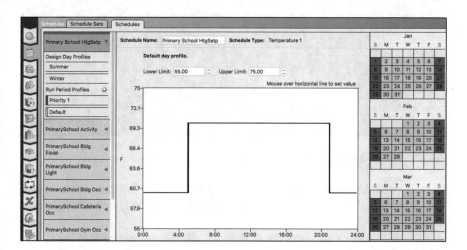

**Fig. 3.55**  Adding PrimarySchool HtgSetp and ClgSetp from the Library

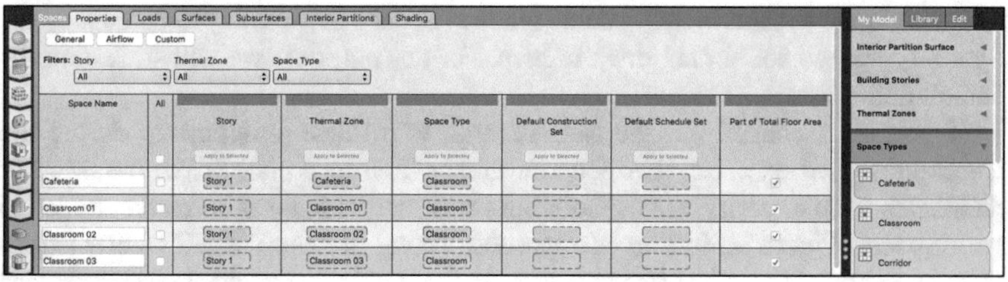

**Fig. 3.56**  All Spaces defaulted to the Classroom Space Type

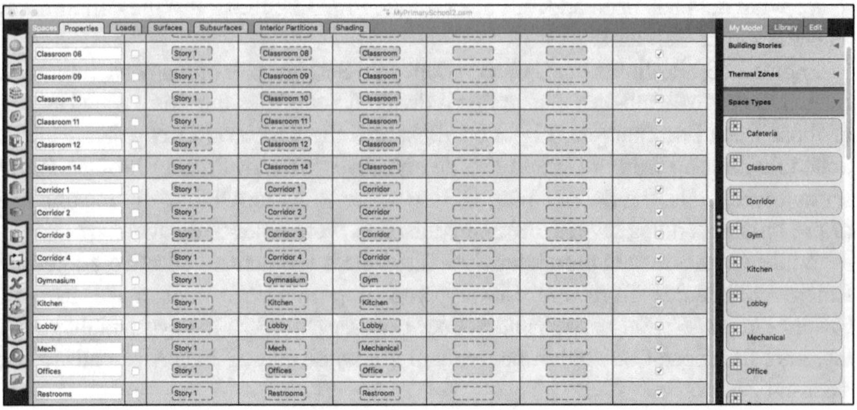

**Fig. 3.57**  Overriding Space Type defaults for other Spaces

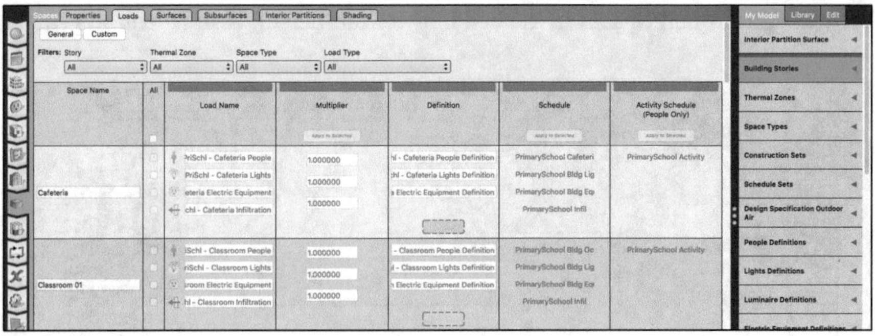

**Fig. 3.58**  Verifying load definitions for Spaces

that's a good thing, but it did incorrectly assign Spaces like the Cafeteria as a Classroom. Drag and drop the correct Space Types onto the appropriate Spaces as shown in Fig. 3.57.

To reinforce our understanding of data inheritance and verify that the correct Loads and Schedules are applied to our Spaces, switch to the [Loads] Sub-Tab. The telltale green text in Fig. 3.58 informs us that Loads and Schedules were inherited from either the Building Default Space Type or the Space Types we manually assigned. If we wished to add specific loads to Spaces that aren't pre-defined by the Space Types, we could drag them in here. For our purposes we will stick with the default definitions.

If you didn't check out the Surface and Subsurface assignments during the Checkpoint Two exercise, this is a convenient time to do so. Select the [Surfaces] or [Subsurfaces] Sub-Tabs to verify that the building Construction Set was correctly applied to your Spaces. Figure 3.59 is an example illustrating that the Default Constructions that were applied to each surface in every Space. Again, we could choose to replace

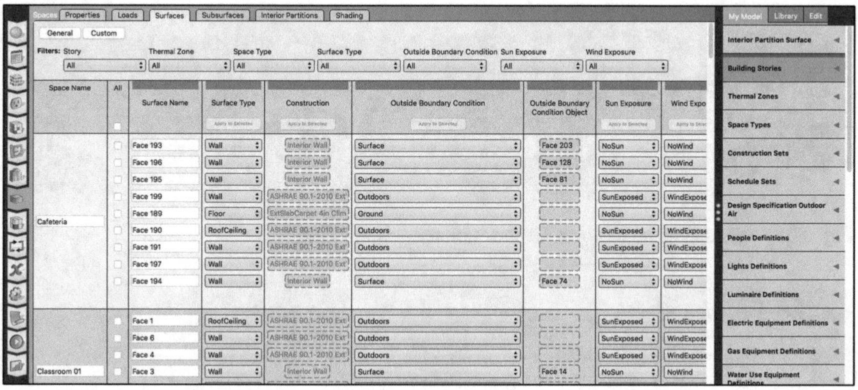

**Fig. 3.59** Verifying Surface Constructions for Spaces

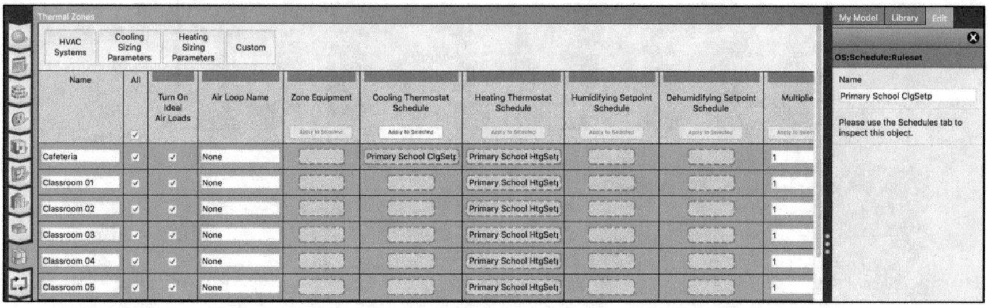

**Fig. 3.60** Using multi-select to assign the cooling Thermostat Schedule to all zones

any of these with a custom construction, but there is no need to do so for our exercise.

The last step in preparing our Model for simulation is to assign our heating and cooling thermostats and turn on Idea Air Loads in each Thermal Zone. Switch to the Zones (▣) Tab shown in Fig. 3.60. Dragging and dropping the Heating and Cooling Schedules onto every Zone in our Model is, pardon the pun, a drag. Fortunately, OpenStudio makes this easier. Use the following steps to assign multiple schedules at once:

1. Click the checkbox under the "All" column. This selects every Thermal Zone in your Model.
2. Drag your Heating Thermostat Schedule onto the correct column for one of the Zones.
3. Click that Schedule to highlight it.
4. Click the ▭▭▭ Button to apply that Schedule to all the selected Thermal Zones.
5. Repeat for the Cooling Thermostat Schedule as shown in Fig. 3.60.
6. Don't forget to check the Ideal Air Load boxes too!

**Fig. 3.61**  Space Type breakdown for the Primary School Model

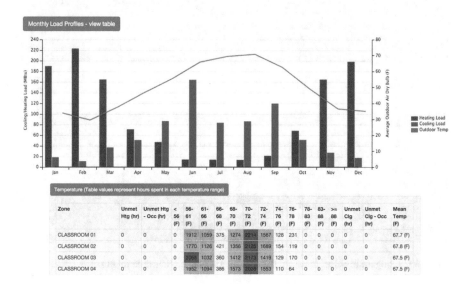

**Fig. 3.62**  HVAC Loads and Zone Conditions for the Primary School Model

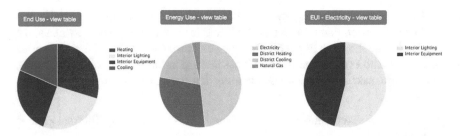

**Fig. 3.63**  Annual end use breakdowns for the Primary School Model

Monthly Overview

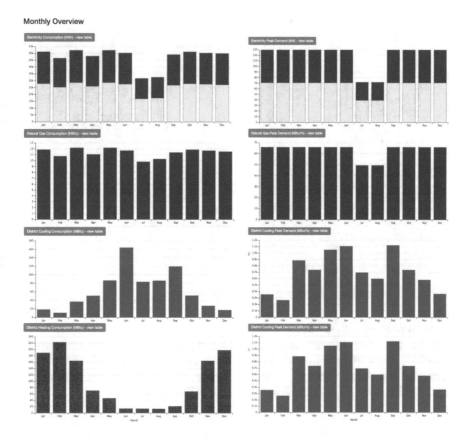

**Fig. 3.64** Monthly end use breakdowns for the Primary School Model

Run the simulation and browse the standard reports. Specific areas of interest are the Space Type Breakdown and Summary sections shown in Fig. 3.61, which allow us to spot check the simulation's end use definitions.

The HVAC Load Profile and Zone Condition sections tell us that the Thermostat Schedules and Ideal Air Loads provided Space conditioning via a fictitious district system (Fig. 3.62). Figures 3.63 and 3.64 show the annual and monthly end use breakdowns for the school.

Lastly, take a look at the time series data for your simulation using DView. Figure 3.65 compares outdoor and Zone temperatures for the Lobby (Red) and Mechanical Room (Orange) over the same time period as in the unconditioned response plots we presented in Fig. 2.56. The lower plot shows the heating and cooling loads from the district system required to achieve the Thermostat target temperatures. Unlike our final results in Checkpoint Three, there is far less evidence of simultaneous heating and cooling, suggesting that our Thermostat Schedules are fairly reasonable.

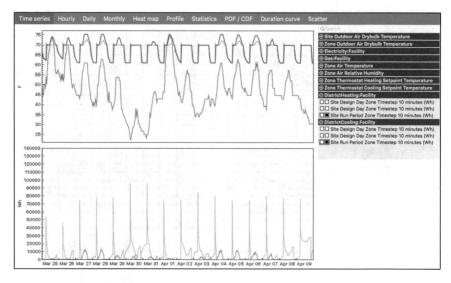

**Fig. 3.65** Time series plots for the Primary School Model

We now have a working knowledge of building Constructions, Loads, Schedules, and the Spaces they support using OpenStudio. We have also learned how to crudely condition those Spaces using Ideal Air Loads to evaluate the reasonableness of our Envelope and Space assumptions. It is now time to turn our attention to modeling how buildings are actually conditioned – using HVAC systems.

## 3.6  Additional Exercises

Use the "Additional Exercises" Model you created in Chap. 2 to continue learning about Space Types:

- Open the Model you created in the Additional Exercises section of Chap. 2,
- Import Space Type definitions into your Library from a similar Model (e.g. School, Office, etc.),
- Assign Space Types to the Spaces in your Model,
- Assign Ideal Air Loads to your Thermal Zones, and
- Run a simulation of your Model.

Review outputs from your model and compare them with results from your previous Checkpoints. Do the end uses in your model make sense? Are the time series plots from your simulations consistent with the Schedules used by your Space Types?

# References

ASHRAE (2013) Handbook fundamentals, ASHRAE, pp 18.3–18.12.

DiLaura D, Houser K, Mistrick R, Stetty G (2011) The lighting handbook, 10th edn. Illuminating Engineering Society (IES). https://www.ies.org/store/lighting-handbooks/lighting-handbook-10th-edition/

http://bigladdersoftware.com/epx/docs/8-7/input-output-reference/group-schedules.html#group-schedules

http://bigladdersoftware.com/epx/docs/8-7/input-output-reference/group-internal-gains-people-lights-other.html#field-activity-level-schedule-name

http://bigladdersoftware.com/epx/docs/8-0/input-output-reference/page-018.html#group-airflow

# Chapter 4
# Introduction to HVAC Systems

## 4.1 Introduction

As we observed in previous Chapter exercises, buildings generally benefit from HVAC systems that are designed to regulate their internal environmental conditions. As the name implies, in addition to heating and cooling, these systems also provide fresh outdoor (ventilation) air to dilute $CO_2$ and other contaminants produced by building occupants, processes, and materials. Modeling HVAC systems correctly is one of the most challenging aspects of energy modeling because of the variety of systems and controls available and the design considerations that drive their selection. The goal of this Chapter is to discuss some of the general concepts needed to understand HVAC system modeling in the context of OpenStudio.

## 4.2 Model Zoning

As described in Chap. 2, OpenStudio Models are divided into Spaces. A Space is a collection of Surfaces and Sub-Surfaces that enclose a volume of air. A Space contains internal loads as described in Chap. 3. In both Chapters, we also briefly touched on the concept of a Thermal Zone. Thermal Zones are served by HVAC systems, and are comprised of one or more Spaces. By extension, a Thermal Zone is a collection of all the Surfaces and Sub-Surfaces that enclose all Zone's Spaces, plus all the Internal Loads contained in those Spaces.

Figure 4.1 shows a building that contains Four Spaces and Two Thermal Zones. One Thermal Zone contains only a single Space, while the other Thermal Zone

The original version of this chapter was revised. A correction to this chapter can be found at https://doi.org/10.1007/978-3-319-77809-9_10

**Electronic Supplementary Material:** The online version of this chapter (https://doi.org/10.1007/978-3-319-77809-9_4) contains supplementary material, which is available to authorized users.

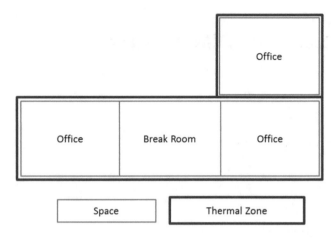

**Fig. 4.1** Two Thermal Zones containing one or more spaces

contains the remaining Three Spaces. Although the Spaces in a Thermal Zone are not required to be adjacent, as described later, this is a best practice.

At every simulation time step, EnergyPlus performs a heat balance calculation for each Thermal Zone. Depending on the thermal boundary conditions discussed in Chap. 2 heat is transferred into or out of a Thermal Zone through its Surfaces. As we learned in Chap. 3, the Thermal Loads within Spaces also transfer heat to the associated Thermal Zone. EnergyPlus generally assumes that all of the air inside of a Thermal Zone is well mixed. This means that all of the heat transferred into the air within a Thermal Zone is instantaneously spread around evenly. There are no hot or cold spots within a Thermal Zone.

Another Model Object closely associated with the Thermal Zone is a Thermostat. As we saw in the previous Chapter Thermostats are associated with temperature setpoints, or targets, and related schedules. HVAC systems attached to Thermal Zones attempt to provide sufficient heat transfer into a Thermal Zone to achieve the target setpoint temperature. Achieving the setpoint takes time based on the size of the HVAC system and attached Thermal Zones, and it is quite possible that target temperatures will not be achieved if the system is undersized for the Thermal Zone Load and boundary conditions. These factors must be considered when "zoning" a building or dividing it into Thermal Zones. Other zoning considerations include the location of Spaces relative to the building façade, variation in heating and cooling setpoints within Spaces, and more. Zoning is in some sense more art than science, but there are a few heuristics that can guide us.

### 4.2.1  Rules of Thumb for Combining Spaces into Thermal Zones

When a Model contains a large number of Spaces, there are many possible ways to combine them into Thermal Zones. While there are no exact rules, the following "rules of thumb" can provide reasonable results:

**Similarity of External Boundary Conditions (Surface Area)**  Think about how much external area (walls, roofs, windows, etc.) each Space has. If the Spaces have similar surface areas, then the well-mixed assumption is probably valid. However, if one Space has many exterior windows and walls, and another Space has none, our engineering judgment and life experience suggests that the Space with no exterior surfaces will be a different temperature than the one with exterior surfaces, especially on very hot or very cold days. Such Spaces should <u>not</u> be grouped together into a common Thermal Zone.

**Similarity of External Boundary Conditions (Timing)**  Consider <u>when</u> external loads are likely to ramp up and down for each Space. For example, if two Spaces are on the top floor of the East side of the building, solar heat transfer will be most intense in the morning, and less so in the late afternoon. Such Spaces are good candidates to combine. Grouping Spaces on East and West façades into a single Thermal Zone may produce less desirable results because they may experience very different exterior boundary conditions at the same time of day.

**Proximity and Connectedness of Spaces**  Think about where in the building the Spaces are located. Although physical adjacency is not a prerequisite for simulation, in reality, it is unlikely that air in two Spaces would be well mixed if the Spaces were not adjacent or not connected by a doorway or corridor that allows air to pass through unimpeded. Proximity may mean adjacency on the same floor or adjacency from one floor to the next. Ask yourself how reasonable is EnergyPlus' "well-mixed" assumption for the collection of Spaces you propose to group together into a single Thermal Zone.

**Size**  Sometimes large Spaces may need to be broken into smaller Spaces if the well-mixed assumption is not reasonable for the entire Space. For example, a large warehouse might seem like a single Space, but is it valid to assume that air heated by exposure to a hot South facing wall will mix with cool air on the North façade? In this case, it may be better to break the warehouse into several smaller Spaces and assign each to a separate Thermal Zone.

**Variation in Internal Loads**  Consider the case of an IT closet with a large number of internal Loads per area and a nearby corridor with little to no Load. Unless there is a direct connection or fan circulating air between these Spaces, it is unlikely that much of the heat from the IT closet will mix into the air in the corridor. These Spaces are candidates for assignment into distinct Thermal Zones.

**Controllability**  Some Spaces may require unique setpoints and schedules. For example, individual rooms in a hotel must offer each guest the ability to set a temperature they find comfortable. In this case, each guest room must be individually zoned.

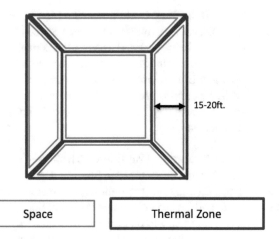

15-20ft.

| Space | | Thermal Zone |

**Fig. 4.2** Core and Perimeter Zoning

## 4.2.2   Zoning During Early Design

It is sometimes desirable to perform simulations after the initial building envelope has been defined but before detailed Spaces have been assigned. In this situation, best practice is to use "Core and Perimeter Zoning." This method divides the building up into a Space for each façade, and then assigns each Space to its own Thermal Zone. Typically, the exterior Spaces are between 15 and 20 feet deep as shown in Fig. 4.2.

## 4.3   HVAC System Types

The HVAC industry offers a large variety of system designs and products. OpenStudio and EnergyPlus are able to model commonly used systems, as well as designs that are far more esoteric. OpenStudio organizes HVAC systems into three basic categories: Zone HVAC, Air Systems, and Plant Systems.

## 4.3.1   Zone HVAC Equipment

Zone HVAC Equipment in OpenStudio refers to a family of components designed to represent a specific, preconfigured, HVAC system that is meant to serve exactly one Thermal Zone. One example of a Zone HVAC component is "Zone HVAC Packaged Terminal Air Conditioner (PTAC)." A common application for a PTAC is in a hotel room, which requires independent air conditioning.

All Zone HVAC types in OpenStudio include a prescribed arrangement of sub-components. In the case of Zone PTAC, these include a fan, a cooling coil, and a heating coil. It is generally not possible to add another sub component, such as a backup heating coil or a humidification device, unless the specific Zone HVAC component was pre-configured to include those sub-components. Just as the component layout is preconfigured for Zone Equipment, so is the associated control logic. Zone Equipment usually includes a limited set of user input fields that can somewhat modify the Equipment's control algorithms. Zone HVAC control logic attempts to follow the Thermal Zone's Thermostat setpoint to the best of the Equipment's ability.

### 4.3.2   Add a Zone HVAC Component

Zone HVAC Equipment such as the PTAC is added to a Thermal Zone using the Application's Thermal Zones (▣) Tab. As with other objects we have used in previous Chapters, simply select the Zone Equipment then drag and drop it onto the desired Thermal Zone as shown in Fig. 4.3. Thermostat Schedules must also be dragged onto the Thermal Zone for the Equipment to function properly.

As with other objects in OpenStudio, we can inspect the properties of the Zone Equipment by clicking on it and looking in the right side of the window (Fig. 4.4). This panel shows all of the detailed properties of the Zone Equipment, as well as the properties of any child components nested inside of it. The PTAC for example has two coils and a fan as children, and the properties of those components are viewable when you inspect the PTAC instance.

Zone Equipment can be useful, but by definition, it is limited to serving single Thermal Zones. The lack of customizability is also a drawback. For greater control of HVAC system configuration, we need to learn about OpenStudio Air Systems.

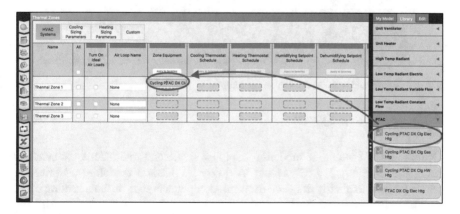

**Fig. 4.3**  Adding a piece of Zone Equipment

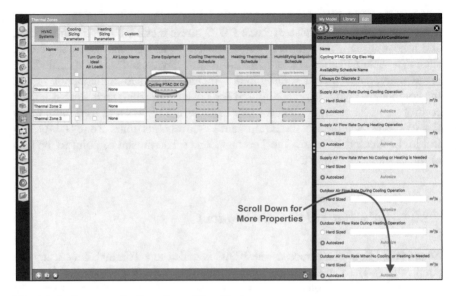

**Fig. 4.4**  Setting properties for a piece of Zone Equipment

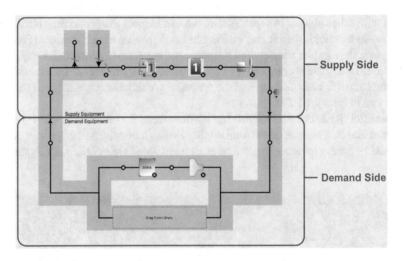

**Fig. 4.5**  Supply and demand sides of an Air Loop

## 4.3.3   Air Loop Systems

OpenStudio is capable of modeling complex single and multi-Zone air handlers using the "AirLoop HVAC" Model. AirLoop HVAC is a modeling container in which you can explicitly drag and drop sub-components such as fans, heating coils, cooling coils, and a wide variety of other equipment. As the name implies AirLoop HVAC is built around the concept of a closed loop with distinct supply and demand sides (Fig. 4.5). The supply side may contain a large variety of fans, coils, heat recovery devices, and outside air systems. The demand side is used to connect Thermal Zones and associated Air Terminal devices.

Compared to Zone Equipment, Air Loops allow for far greater configurability. The user has nearly complete[1] flexibility in the selection and placement of subcomponents, as well as more customizable control options. Additionally, Air Loop HVAC can be attached to one or many Thermal Zones using a variety of Zone Terminal units.

When modeling single Thermal Zone systems, OpenStudio modelers often have a choice between using the Air Loop Model or alternately using one of the Zone HVAC Objects pre-built for a specific purpose. In fact, there is overlap in the capabilities of the two approaches. Because of the configurability that Air Loop offers, it is capable of modeling many practical single Thermal Zone air-based systems. That said there are single Thermal Zone systems that are only possible to model properly in OpenStudio as Zone Equipment. In general, it is recommended to use Zone Equipment when there is a suitable model available and the Equipment serves a single Thermal Zone.

Lastly note that OpenStudio allows for multiple pieces of Zone Equipment to be attached to a single Thermal Zone, however Thermal Zones may only have one Air Loop Connection. A common scenario is to use Zone Equipment to condition a single Thermal Zone, for instance the "Zone HVAC Four Pipe Fan Coil" Object and reserve the more configurable Air Loop to model a dedicated outside air system (DOAS). These options will become clearer in subsequent sections and exercises.

### 4.3.4  Plant Loop Systems

Liquids are frequently used as a heat transfer medium in HVAC systems. All liquid based systems, including chilled and hot water plants, condenser systems, and potable hot water, are modeled using the Plant Loop component Model. Like Air Loops, Plant Loops are closed loops incorporating the concept of a supply side and demand side. The supply side of a Plant Loop typically contains heat producing or extracting components such as chillers, boilers, and cooling towers. The demand side is typically composed of consumers, such as chilled and hot water coils. Thus, the demand side of Plant Loops are frequently connected to the supply side of Air Loops as shown in Fig. 4.6.

Plant Loops allow some limited flow path branching. Specifically, splitters and mixers paired on the supply side enables configurations like gangs of chillers. Plant Loops also allow a single splitter and mixer to be paired up on the demand side,

---

[1] OpenStudio does impose some limits on placement of components relative to the supply and demand side of Air Loops to prevent the user from accidentally creating Models that won't simulate in EnergyPlus.

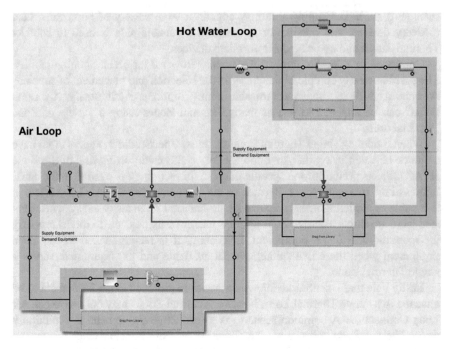

**Fig. 4.6**  A typical Plant and Air Loop relationship

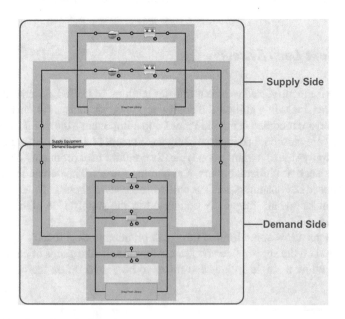

**Fig. 4.7**  Plant Loop branch examples

typically with parallel branches serving various plant loads such as water coils. Figure 4.7 illustrates common Plant branching scenarios.

One constraint on Plant Loop topologies is that it is not possible to have a splitter follow another splitter to form a sub-branch (Fig. 4.8). This is a limitation of the underlying EnergyPlus simulation engine. In practice, Plant Loop branching options in OpenStudio are adequate to model the majority of real world systems. Nevertheless, it is helpful to keep the concepts of supply and demand side and constrained branching, in mind when designing plant systems in OpenStudio.

## 4.4   HVAC System Templates

Compared to Zone Equipment, Air and Plant Loop systems are more complex to create from scratch. We will discuss custom built Loop-based systems in Chap. 5, but for now let's take a look at using "templates" that OpenStudio provides to quickly construct commonly used HVAC systems. These templates roughly correspond to a subset of the ASHRAE 90.1 Appendix G system types.[2] Template systems may also be used as a starting point and modified as needed.

Consider ASHRAE 90.1 Appendix G system type seven. System seven is a built-up system that has a multi-Zone, variable air volume (VAV), central air handler (Fig. 4.9). There are chilled and hot water coils in the air handler, which are fed by chilled and hot water plants respectively. The chilled water plant is cooled by a

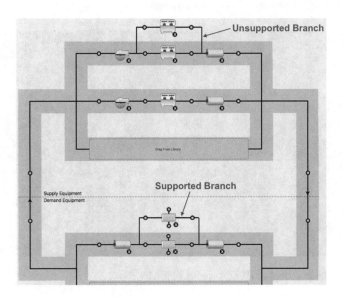

**Fig. 4.8**  An unsupported Plant Loop topology

---

[2] ASHRAE (2016).

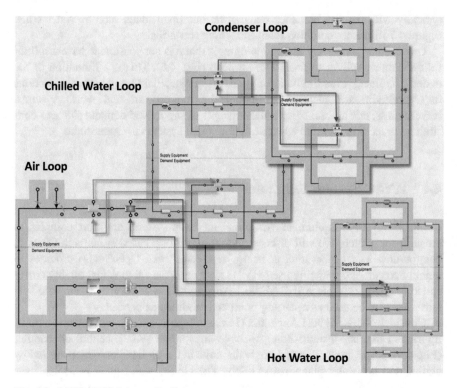

**Fig. 4.9**   ASHRAE 90.1 Appendix G system type seven

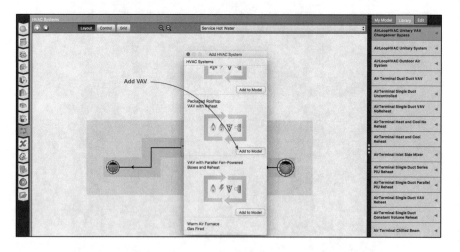

**Fig. 4.10**   Adding a template HVAC system to a Model

water-cooled chiller, which is rejecting heat to another Plant Loop serving as the condenser system, which rejects heat via a cooling tower. In total, there are three Plant Loop instances and one Air Loop HVAC instance used to model system type seven in OpenStudio. It is possible to build up these systems from scratch using Air and Plant Loops and HVAC components, but it is far easier to start from an OpenStudio template.

### 4.4.1   Add a Template Air Loop HVAC System

Template HVAC systems are added using the HVAC (▣) Tab in the Application. In the upper left corner of the user interface, there is a ▣ Button to select and add template systems. Clicking this Button opens an "Add HVAC System" dialog as shown in Fig. 4.10. Scrolling through the options reveals a number of common system configurations. ASHRAE system type seven most closely resembles the "Packaged Rooftop VAV with Reheat" template. Clicking the ⬚Add to Model⬚ Button adds all of the necessary Plant and Air Loops along with the requisite sub-components to the Model.

This particular template adds four new Loops to the Model including one Air Loop representing the central air handler, a hot water Plant Loop, a chilled water Plant Loop, and a condenser Plant Loop. The template is preconfigured with components linking the Loops together. After the template has been added, the system selector field near the top of the HVAC (▣) Tab changes from "Service Hot Water" to "VAV with Reheat," displaying the Air Loop representing the central air handler (Fig. 4.11). This field may be used to navigate between Loops and other systems that may be present in the Model.

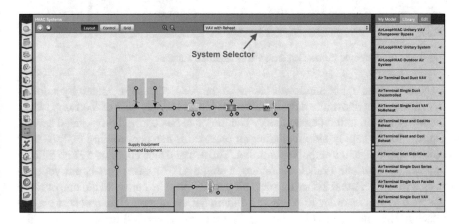

**Fig. 4.11**  Air handler included in Packaged Rooftop VAV with reheat templatesystem

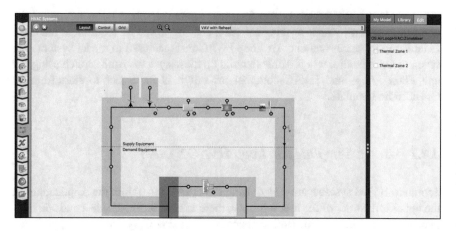

**Fig. 4.12**  Connecting Thermal Zones to an Air handler

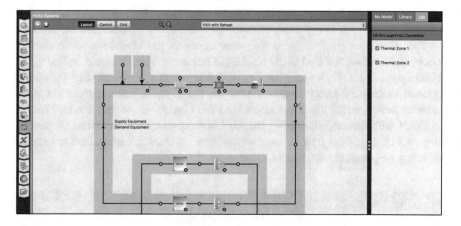

**Fig. 4.13**  Two Thermal Zones Connected to a Single Air Handler

Note that the upper and lower halves of the Air Loop are separated by a dotted line. The region above the dotted line represents the supply side of the Loop, while the lower region is the demand side. The template has conveniently added several Objects to the supply side including an outside air handler, cooling coil, heating coil, supply fan, and a setpoint controller. The demand side contains a Zone Splitter and Mixer, with one branch containing a single Air Terminal that is not yet connected to any Thermal Zone. The new air handler must be connected to one or more Thermal Zones in order to produce a functional simulation. The quickest way to attach a Thermal Zone is by clicking on either the Zone Splitter or Mixer on the demand side, then selecting the Thermal Zones to attach using the inspector interface on the right as shown in Fig. 4.12.

After checking Thermal Zones 1 and 2, the Air Loop updates to reflect two demand side branches, each containing a VAV Air Terminal serving a Thermal Zone as shown in Fig. 4.13.

It is also possible to add a Thermal Zone by selecting a Thermal Zone from the My Model Sub-Tab and dragging it either to the region labeled "Drag From Library," or on a Node immediately after an Air Terminal.

Chapter 5 will go into greater detail regarding the OpenStudio HVAC interface, but let's take a moment to describe a few rudimentary features at this point. Most importantly, all of the components shown in the HVAC interface may be inspected. Clicking an icon will usually present an Object's properties in the right-hand side of the window. As with Zone Equipment, these include all Object parameters as well as the properties of any children the component may include. For example, clicking on the VAV Air Terminal brings up the Air Terminal properties, as well as the properties for the hot water heating coil contained inside the terminal as shown in Fig. 4.14. Some Objects offer more options next to the ▨ Sub-Tab, which contains most general Properties. These may include a ▨ Sub-Tab, which summarizes component "linkages" with other Loops, and a ▨ Sub-Tab, which contains controls-related options. Additional Sub-Tabs will appear when the selected Object supports them.

Nodes are the little ▰ symbols scattered around a Loop and are used to separate all of the Objects contained within Loops. This is an important concept in OpenStudio, because Nodes represent points where physical fluid properties, such as temperature, humidity ratio, and flow rate are known and can be reported. Nodes may also be associated with control points. In Chap. 5 we will discuss how control setpoints may be applied to Nodes in order to drive supply components to follow constant or scheduled performance targets.

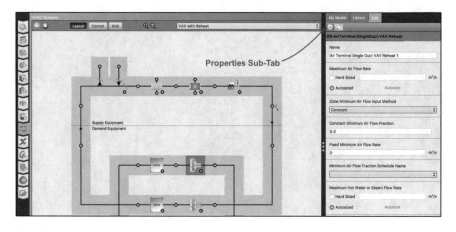

**Fig. 4.14** Examining the Properties of an Air Terminal

Managing Nodes and Connections between Objects can be a very tedious and error prone task when using EnergyPlus directly, however OpenStudio's Object model helps take care of this tedious work automatically. Objects may be removed by clicking on their corresponding ⊗ Button. New Objects may be added by dragging them from the ▆▆▆ on the right side of the interface and dropping them on to existing Nodes. In either case, OpenStudio adds and remove Nodes as needed so that the user does not need to worry about maintaining Node Connections manually.

Air and Plant Loops are shown individually in OpenStudio and may be browsed using the system selector field at the top of the window. OpenStudio also makes it easy to navigate between interconnected Loops. In this example, the air handler is connected to chilled- and hot-water plants through chilled- and hot-water coils respectively. To navigate to these associated Plant Loops, simply click on either of the smaller Node icons directly above or below a coil as shown in Fig. 4.15.

The interface immediately switches to the correct Plant Loop shown in Fig. 4.16. The chilled water Plant Loop contains the same coil shown on the supply side of the Air Loop. However, from the Plant Loop's perspective the coil appears on the demand side of the chilled water Plant. The chilled water coil's Node links may be used to navigate back to the Air Loop from the Plant Loop.

The Plant Loop contains a pump feeding a single chiller on the supply side of the chilled water plant. Note that the chiller contains its own Nodes above and below the chiller icon that may be used to navigate to the chiller's associated condenser

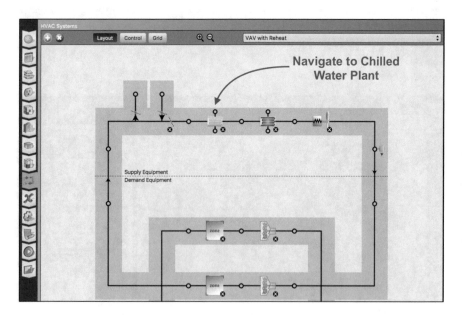

**Fig. 4.15** Clicking a Chilled Water Coil Node to Navigate to the Associated Plant Loop

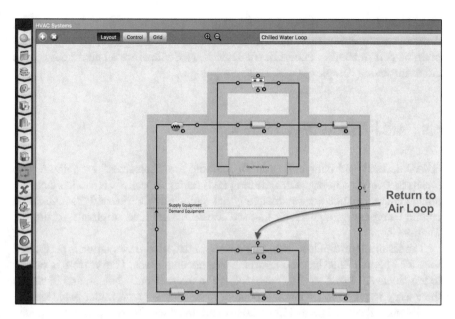

**Fig. 4.16**   Plant Loop Serving the Air Handler Chilled Water Coil

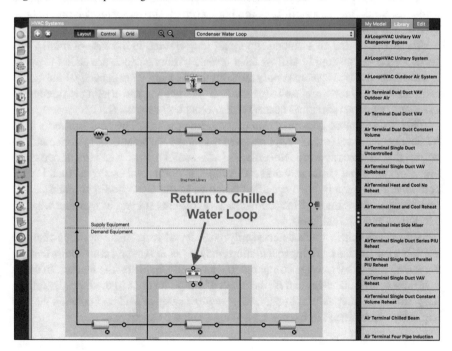

**Fig. 4.17**   Condenser Loop Serving the Chilled Water Plant

Plant Loop (Fig. 4.17). In this case, the chiller is water-cooled and is modeled as a demand component on the condenser system. Since one Loop's supply objects may often be part of another Plant Loop's demand, Node links are a helpful navigation aid in traversing complex HVAC system topologies.

## 4.5  Auto-Sizing HVAC Systems

HVAC systems and components in OpenStudio are "Autosized" by default. This means that equipment flow rates, heating and cooling capacities, and other capacity related characteristics are automatically determined by the EnergyPlus simulation engine using sizing algorithms that are driven by the load originating from the Thermal Zones.

The sizing algorithm is a cascading process. First, a load calculation is performed for each Thermal Zone based on extreme weather conditions. This process is similar to the "Ideal Air Load" method we explored in the previous Chapter, and identifies the energy required to maintain a cooling or heating setpoint under peak loading conditions. All of the Thermal Zone Loads associated with HVAC systems are added together to arrive at a total system Load. Individual component sizes are computed using that total Load as a starting point. For forced air-driven system involving fans, system sizing begins by calculating a supply airflow rate based on the Thermal Zone Load calculation and an assumed supply air temperature. In the case of multi-Zone systems, the design supply airflow rates for each Thermal Zone are added together to determine the total system supply airflow rate and supply fan capacity. Finally, the computed airflow rates are used along with design heating and cooling temperatures in order to size the individual heating and cooling coil capacities.

This is an extreme generalization of the sizing process. The actual process used by EnergyPlus is sufficiently sophisticated[3] to handle the breadth of system configurations that may be modeled. Nevertheless, it is useful to have a general understanding about how the process works. The most important takeaway is that HVAC components ultimately have specific capacities in the OpenStudio simulation that are determined by EnergyPlus' auto-sizing algorithms subject to extreme weather conditions and internal loads.

Auto-sized values do not necessarily correspond to realistic systems that are available for purchase. Sizing calculations may result in system components that are unrealistically small or large compared to what are available on the market. In many cases, these artificial sizes can be the result of simplifications in the Model, particularly around Zoning assumptions. Lastly, in most cases HVAC components may be explicitly "hard-sized" to fixed values using available Object parameters.

---

[3] Refer to the EnergyPlus Engineering Reference for discussion of the sizing process. http://bigladdersoftware.com/epx/docs/8-7/engineering-reference/.

### 4.5.1   Design Day Files

As mentioned in Chap. 2, DDY weather files describe the extreme weather conditions used for auto-sizing. Any OpenStudio Model that contains auto-sized HVAC systems must have at least one Design Day input in order to simulate without error. In practice, almost all OpenStudio Models that contain HVAC have at least one auto-sized property and therefore must have Design Day inputs defined. Recall from Chap. 2 that design day DDY files are imported using the Site (◼) Tab (Fig. 4.18).

A DDY file typically contains several different design points for heating and cooling. OpenStudio imports only the subset corresponding to the 0.4% Summer design day point, and the 99.6% Winter design day. This means that in the Summer the design temperature will only be exceeded 0.4% of the time, and in the Winter the temperature will remain above the design temperature for 99.6% of the hours in the year. After importing a DDY file, the Site (◼) Tab updates to show the imported design days (Fig. 4.19). It is possible to add and remove design days manually, using the ◼ and ◼ Buttons.

## 4.6   Checkpoint Five: Adding a Template HVAC System to a School Model

In this exercise, we will remove the Ideal Air Loads we added to our primary school Model and replace them with a proper HVAC system. To begin the exercise:

1. Open your Primary School Model (or a copy of it) from Checkpoint Four.

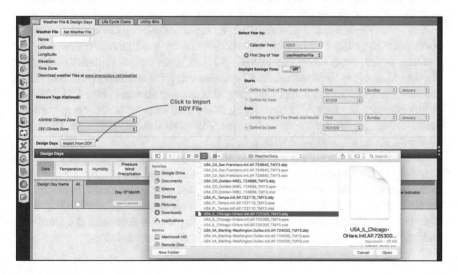

**Fig. 4.18** Adding Design Days from a DDY File

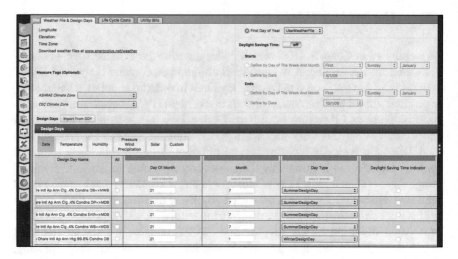

**Fig. 4.19** Design Days Added to an OpenStudio Model

2. Use Save As to make a copy of the Model called *MyPrimarySchoolHVAC.osm*.
3. Navigate to the (■) Tab and import design days from *USA_CO_Golden-NREL.724666_TMY3.ddy*.
4. Navigate to the Thermal Zones (■) Tab, check all of the Zones, and delete them with the ■ Button since we are going to rezone the Model.
5. Save your Model.

Begin by Zoning the school Model as shown in Fig. 4.20. The Model should include four Thermal Zones named Thermal Zone 1, Thermal Zone 2, Thermal Zone 3, and Thermal Zone 4. Feel free to perform this task with the floor plan editor or in the Thermal Zones (■) Tab, whichever you are most comfortable using.

Save your work. We also recommend you make a spare copy of the Model at this point, perhaps called *MyPrimarySchoolZoned.osm* for use in the next Chapter.

Once saved, proceed to the HVAC (■) Tab. Use the ■ Button to add four "Packaged Rooftop VAV with Reheat" systems using the available templates. Then add one Thermal Zone to each of the four Air Loops as shown in Fig. 4.21. You can switch between the four Air Loops using the system selector at the top of the window. By default, the Air Loops are named VAV with Reheat, VAV with Reheat 1, VAV with Reheat 2, and VAV with Reheat 3.

Take a few moments to explore some of the Objects that have been automatically added to the Air Loops for you. You may also wish to use the available Node links to examine the Plant Loops connected to the hot and cold coils as well. For example, the hot water Plant loop for Thermal Zone One's air handler is shown in Fig. 4.22. Note that this Plant Loop has <u>two</u> hot water coils on the demand side. Does this make sense? Can you locate the second coil located in the Air Loop?

**Fig. 4.20** Primary School
Model with Four Thermal
Zones

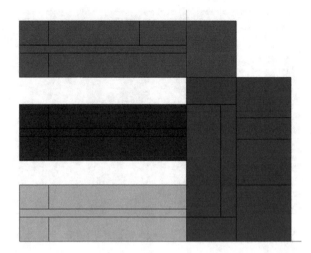

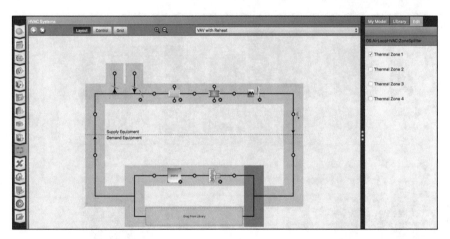

**Fig. 4.21** Adding Thermal Zone One to an Air Handler

Run the simulation and examine the OpenStudio Results report. We've looked at
the monthly overview and Thermal Zone overview sections of this report in Chap.
3. The district system heating and cooling energy used by the Ideal Air Loads in
Checkpoint Four have now been replaced with increased electricity and gas use
(Fig. 4.23). The Thermal Zone condition portion of the report shown in Fig. 4.24
indicates that the system is doing a reasonably good job of maintaining the setpoint
temperatures throughout the year.

Along with the Thermal Zone overview shown in Fig. 4.25, additional sections
of the report are now populated with details related to our HVAC systems (Fig. 4.26).

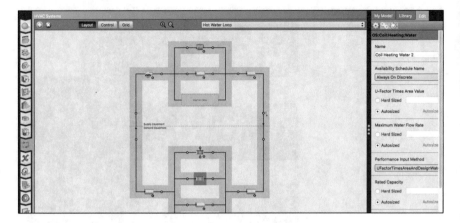

**Fig. 4.22** Hot Water Loop Serving Air Handler for Thermal Zone One

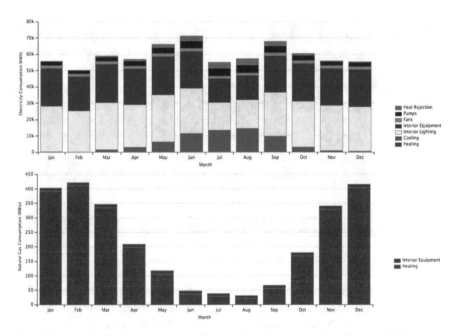

**Fig. 4.23** Monthly Electricity and Gas Consumption for School with Four Packaged VAV Units

| Zone | Unmet Htg (hr) | Unmet Htg - Occ (hr) | < 56 (F) | 56-61 (F) | 61-66 (F) | 66-68 (F) | 68-70 (F) | 70-72 (F) | 72-74 (F) | 74-76 (F) | 76-78 (F) | 78-83 (F) | 83-88 (F) | >= 88 (F) | Unmet Clg (hr) | Unmet Clg - Occ (hr) | Mean Temp (F) |
|---|---|---|---|---|---|---|---|---|---|---|---|---|---|---|---|---|---|
| THERMAL ZONE 1 | 105 | 0 | 0 | 1239 | 1664 | 354 | 2559 | 747 | 778 | 1419 | 0 | 0 | 0 | 0 | 0 | 0 | 68.4 (F) |
| THERMAL ZONE 2 | 105 | 0 | 0 | 1229 | 1667 | 357 | 2549 | 755 | 777 | 1426 | 0 | 0 | 0 | 0 | 0 | 0 | 68.4 (F) |
| THERMAL ZONE 3 | 104 | 0 | 0 | 1220 | 1678 | 358 | 2506 | 727 | 779 | 1492 | 0 | 0 | 0 | 0 | 0 | 0 | 68.4 (F) |
| THERMAL ZONE 4 | 123 | 0 | 0 | 954 | 1848 | 442 | 2536 | 761 | 886 | 1333 | 0 | 0 | 0 | 0 | 0 | 0 | 68.5 (F) |

| Zone | < 30 (%) | 30-35 (%) | 35-40 (%) | 40-45 (%) | 45-50 (%) | 50-55 (%) | 55-60 (%) | 60-65 (%) | 65-70 (%) | 70-75 (%) | 75-80 (%) | >= 80 (%) | Mean Relative Humidity (%) |
|---|---|---|---|---|---|---|---|---|---|---|---|---|---|---|
| THERMAL ZONE 1 | 5032 | 723 | 695 | 668 | 532 | 413 | 350 | 153 | 122 | 65 | 7 | 0 | 30.4 (%) |
| THERMAL ZONE 2 | 5040 | 718 | 699 | 669 | 528 | 415 | 347 | 150 | 124 | 63 | 7 | 0 | 30.4 (%) |
| THERMAL ZONE 3 | 5034 | 731 | 699 | 673 | 516 | 425 | 339 | 151 | 124 | 61 | 7 | 0 | 30.3 (%) |
| THERMAL ZONE 4 | 5142 | 756 | 640 | 656 | 502 | 459 | 287 | 149 | 125 | 44 | 0 | 0 | 29.6 (%) |

**Fig. 4.24**  Zone Conditions for School with Four Packaged VAV Units

| | Area (ft^2) | Conditioned (Y/N) | Part of Total Floor Area (Y/N) | Volume (ft^3) | Multiplier | Above Ground Gross Wall Area (ft^2) | Underground Gross Wall Area (ft^2) | Window Glass Area (ft^2) | Lighting (W/ft^2) | People (ft^2/person) | Plug and Process (W/ft^2) |
|---|---|---|---|---|---|---|---|---|---|---|---|
| THERMAL ZONE 1 | 14280.05 | Yes | Yes | 185640.03 | 1.00 | 6344.03 | 0.0 | 1903.17 | 1.15 | 47.25 | 0.91 |
| THERMAL ZONE 2 | 14280.05 | Yes | Yes | 185640.03 | 1.00 | 6344.03 | 0.0 | 1903.17 | 1.15 | 47.25 | 0.91 |
| THERMAL ZONE 3 | 18495.95 | Yes | Yes | 240448.03 | 1.00 | 8033.97 | 0.0 | 2410.15 | 1.17 | 46.18 | 0.93 |
| THERMAL ZONE 4 | 26000.01 | Yes | Yes | 337999.85 | 1.00 | 6006.05 | 0.0 | 1801.77 | 0.91 | 56.73 | 5.01 |
| Total | 73055.95 | | | 949727.95 | | 26727.97 | 0.0 | 8018.36 | 1.07 | 49.94 | 2.38 |
| Conditioned Total | 73055.95 | | | 949727.95 | | 26727.97 | 0.0 | 8018.36 | 1.07 | 49.94 | 2.38 |

**Fig. 4.25**  Zone Overview for School Divided into Four Thermal Zones

The EnergyPlus standard report is also full of insightful information. For example, the Comfort and Setpoint Not Met Summary shown in Fig. 4.27 corroborates what we noted in Fig. 4.25 above.

Also, note a link in the report's table of contents summarizing "component sizing." These are results from EnergyPlus' auto-sizing algorithm and appear in Fig. 4.28 below. The values in these tables correspond to the calculated Thermal Zone Loads, as well as information pertaining to the capacities of each HVAC component found in the Model. If components were hard-sized then the component sizing summary will echo out the Equipment capacities provided by the user.

As we will see in Chap. 5, we have just scratched the surface of OpenStudio and EnergyPlus capability to model HVAC systems.

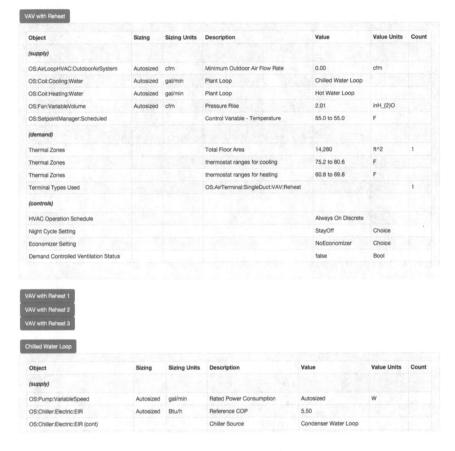

**VAV with Reheat**

| Object | Sizing | Sizing Units | Description | Value | Value Units | Count |
|---|---|---|---|---|---|---|
| *(supply)* | | | | | | |
| OS:AirLoopHVAC:OutdoorAirSystem | Autosized | cfm | Minimum Outdoor Air Flow Rate | 0.00 | cfm | |
| OS:Coil:Cooling:Water | Autosized | gal/min | Plant Loop | Chilled Water Loop | | |
| OS:Coil:Heating:Water | Autosized | gal/min | Plant Loop | Hot Water Loop | | |
| OS:Fan:VariableVolume | Autosized | cfm | Pressure Rise | 2.01 | inH_{2}O | |
| OS:SetpointManager:Scheduled | | | Control Variable - Temperature | 55.0 to 55.0 | F | |
| *(demand)* | | | | | | |
| Thermal Zones | | | Total Floor Area | 14,280 | ft^2 | 1 |
| Thermal Zones | | | thermostat ranges for cooling | 75.2 to 80.6 | F | |
| Thermal Zones | | | thermostat ranges for heating | 60.8 to 69.8 | F | |
| Terminal Types Used | | | OS:AirTerminal:SingleDuct:VAV:Reheat | | | 1 |
| *(controls)* | | | | | | |
| HVAC Operation Schedule | | | | Always On Discrete | | |
| Night Cycle Setting | | | | StayOff | Choice | |
| Economizer Setting | | | | NoEconomizer | Choice | |
| Demand Controlled Ventilation Status | | | | false | Bool | |

**VAV with Reheat 1**
**VAV with Reheat 2**
**VAV with Reheat 3**

**Chilled Water Loop**

| Object | Sizing | Sizing Units | Description | Value | Value Units | Count |
|---|---|---|---|---|---|---|
| *(supply)* | | | | | | |
| OS:Pump:VariableSpeed | Autosized | gal/min | Rated Power Consumption | Autosized | W | |
| OS:Chiller:Electric:EIR | Autosized | Btu/h | Reference COP | 5.50 | | |
| OS:Chiller:Electric:EIR (cont) | | | Chiller Source | Condenser Water Loop | | |

**Fig. 4.26** Air and Plant Loop Detail for the Revised School Model

## 4.7　Additional Exercises

1. Recommended additional exercises involving the Checkpoint Five Model include further study of template HVAC systems and the impact of zoning on system performance.

   - **Be sure to keep a "clean" copy of the Checkpoint Five model before proceeding with these activities.**
   - Replacement of the "Packaged Rooftop VAV with Reheat" template system

   - Remove the existing air and plant loops and replace them with new template systems
   - Reattach the original zones to the new air loop(s)
   - Visually inspect the resulting air and plant loops noting the differences
   - Run the new models and compare their performance with Checkpoint Five.

## Setpoint Not Met Criteria

|  | Degrees [deltaC] |
|---|---|
| Tolerance for Zone Heating Setpoint Not Met Time | 0.20 |
| Tolerance for Zone Cooling Setpoint Not Met Time | 0.20 |

## Comfort and Setpoint Not Met Summary

|  | Facility [Hours] |
|---|---|
| Time Setpoint Not Met During Occupied Heating | 0.00 |
| Time Setpoint Not Met During Occupied Cooling | 0.00 |
| Time Not Comfortable Based on Simple ASHRAE 55-2004 | 1225.67 |

**Fig. 4.27**  EnergyPlus Unmet Hours Report

AirTerminal:SingleDuct:VAV:Reheat

|  | Design Size Maximum Air Flow Rate [m3/s] | Design Size Constant Minimum Air Flow Fraction | User-Specified Constant Minimum Air Flow Fraction | Design Size Minimum Air Flow Rate [m3/s] | Design Size Maximum Flow per Zone Floor Area during Reheat [m3/s-m2] | Design Size Maximum Flow Fraction during Reheat [] | Design Size Maximum Reheat Water Flow Rate [m3/s] | Design Size Reheat Coil Sizing Air Volume Flow Rate [m3/s] | Design Size Reheat Coil Sizing Inlet Air Temperature [C] | Design Size Reheat Coil Sizing Inlet Air Humidity Ratio [kgWater/kgDryAir] |
|---|---|---|---|---|---|---|---|---|---|---|
| AIR TERMINAL SINGLE DUCT VAV REHEAT 1 | 4.53 | 0.223283 | 0.300000 | 1.36 | 0.001024 | 0.300000 | 0.001848 | 3.13 | 12.80 | 0.008000 |
| AIR TERMINAL SINGLE DUCT VAV REHEAT 2 | 4.55 | 0.221990 | 0.300000 | 1.37 | 0.001030 | 0.300000 | 0.001845 | 3.12 | 12.80 | 0.008000 |
| AIR TERMINAL SINGLE DUCT VAV REHEAT 3 | 5.96 | 0.219805 | 0.300000 | 1.79 | 0.001040 | 0.300000 | 0.002340 | 3.96 | 12.80 | 0.008000 |
| AIR TERMINAL SINGLE DUCT VAV REHEAT 4 | 7.46 | 0.246697 | 0.300000 | 2.24 | 0.000927 | 0.300000 | 0.002555 | 4.32 | 12.80 | 0.008000 |

*User-Specified values were used. Design Size values were used if no User-Specified values were provided. Design Size values may be derived from alternate User-Specified values.*

Coil:Heating:Water

|  | Design Size Rated Capacity [W] | Design Size Maximum Water Flow Rate [m3/s] | Design Size U-Factor Times Area Value [W/K] |
|---|---|---|---|
| COIL HEATING WATER 2 | 85074.12 | 0.001848 | 1742.71 |
| COIL HEATING WATER 4 | 84936.53 | 0.001845 | 1739.89 |
| COIL HEATING WATER 6 | 107710.89 | 0.002340 | 2206.41 |
| COIL HEATING WATER 8 | 117587.12 | 0.002555 | 2408.68 |
| COIL HEATING WATER 1 | 41623.17 | 0.000904 | 540.15 |
| COIL HEATING WATER 3 | 41865.55 | 0.000910 | 543.30 |
| COIL HEATING WATER 5 | 54764.80 | 0.001190 | 710.69 |
| COIL HEATING WATER 7 | 68591.72 | 0.001490 | 890.13 |

*User-Specified values were used. Design Size values were used if no User-Specified values were provided.*

AirLoopHVAC

|  | Design Supply Air Flow Rate [m3/s] |
|---|---|
| VAV WITH REHEAT | 4.53 |
| VAV WITH REHEAT 1 | 4.55 |
| VAV WITH REHEAT 2 | 5.96 |
| VAV WITH REHEAT 3 | 7.46 |

*User-Specified values were used. Design Size values were used if no User-Specified values were provided.*

**Fig. 4.28**  EnergyPlus Auto-Sizing Report

- Rezoning of the Model
- Modify the Thermal Zones from the original Model and compare the performance difference with:
  - Fewer Thermal Zones,
  - More Thermal Zones, and
  - Different Space groupings in the Thermal Zones
- What are the impacts on energy usage?
- What happens to system sizing?
- Do the number of unmet hours change?

2. Use the "Additional Exercises" Model you created in Chap. 3.

- Attempt to zone your Model as accurately as possible by:
- Interviewing your building's facility manager to understand how the building is zoned
- Identifying individual thermostats located within your building's spaces and grouping Spaces within the Model based on their locations.
- Apply one or more template HVAC systems to your Model
- Compare the performance of the systems you've selected
- Modify your Thermal Zoning assumptions and see how performance changes

# References

ANSI/ASHRAE/IES Standard 90.1-2016 energy standard for buildings except low-rise residential buildings, ASHRAE, 2016
http://bigladdersoftware.com/epx/docs/8-7/engineering-reference

# Chapter 5
# Advanced HVAC Topics

## 5.1 Introduction

As described in Chap. 4, there are three main categories of HVAC Equipment in EnergyPlus: Plant Loops, Air Loops, and Zone Equipment. This chapter goes into more detail for each of these categories, describing their configuration, sizing, control, and operation. HVAC is a complex topic, and for that reason this chapter only covers the most critical concepts and topics. The authors suggest reading the EnergyPlus Engineering Reference as a supplement to this chapter to learn more.

## 5.2 Air Loops

Recall that Air Loop are a series of Objects representing HVAC systems that conditions air, which is provided to Thermal Zones for heating, cooling, and ventilation.

### 5.2.1 Air Loop Configuration

In OpenStudio, an Air Loop is represented by the diagram shown in Fig. 5.1. Air always flows in the directions shown by the red arrows. An Air Loop, assuming it has an outdoor air system with an outdoor air intake and exhaust, is an open system, where some amount of the return air is exhausted and replaced with fresh outdoor air for ventilation.

---

The original version of this chapter was revised. A correction to this chapter can be found at https://doi.org/10.1007/978-3-319-77809-9_10

**Electronic Supplementary Material:** The online version of this chapter (https://doi.org/10.1007/978-3-319-77809-9_5) contains supplementary material, which is available to authorized users.

© Springer International Publishing AG, part of Springer Nature 2018                127
L. Brackney et al., *Building Energy Modeling with OpenStudio*,
https://doi.org/10.1007/978-3-319-77809-9_5

**Fig. 5.1** Flow direction in
an empty Air Loop

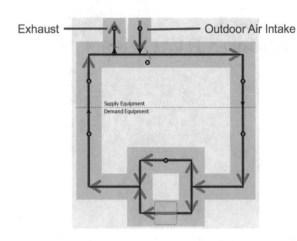

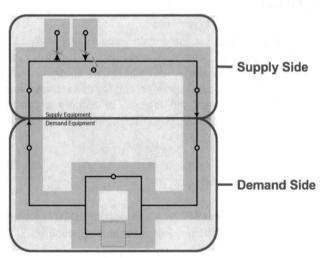

**Fig. 5.2** Supply and demand sides of an empty Air Loop

As shown in Fig. 5.2, an Air Loop can first be broken into two pieces, supply and demand sides.

### 5.2.1.1  Air Loop Supply Side

The supply side of an Air Loop is responsible for:

- Consuming air returning from Thermal Zones via the supply inlet Node,
- Exhausting some of that air to an outdoor air system,
- Consuming replacement air via an outdoor air system,
- Heating, cooling, and humidifying it to the correct supply air conditions, and
- Supplying it to the Thermal Zones via a supply outlet Node.

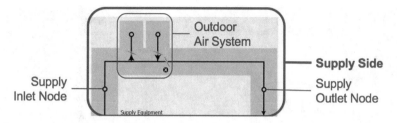

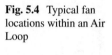

**Fig. 5.3**  Empty supply side Air Loop detail

**Fig. 5.4**  Typical fan
locations within an Air
Loop

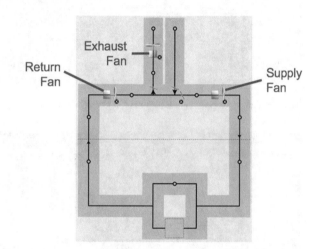

An empty supply side portion of an Air Loop is shown in Fig. 5.3. Supply side
components may be added to Loop between the supply inlet and outlet Nodes.
Outdoor air system components may be added just below the outdoor air exhaust
and intake Nodes.

In EnergyPlus, an Air Loop must contain at least one fan on the supply side but
may include more than one. Figure 5.4 illustrates three typical locations for a fan. A
fan positioned before the outdoor air system is commonly called a return fan. A fan
positioned adjacent to the outdoor air system exhaust Node is frequently referred to
as an exhaust or relief fan. A fan positioned downstream of the outdoor air system
is often called a supply fan.

In EnergyPlus, unlike in real HVAC systems, fans do not drive airflow around the
loop. Instead fan Objects utilize the amount of air flowing through them along with
prescribed pressure drop and efficiency curves to calculate the energy that would be
consumed to move that much air. Supply and return fans must accommodate the full
flow rate of the Air Loop, while exhaust fans only experience the exhaust air flow
rate. This may seem an odd role for EnergyPlus fan Objects but will become more
clear when we discuss how air flow rates are calculated in Sects. 5.2.2 and 5.2.4.

Besides fans, common supply side Equipment includes:

- Outdoor air systems required for systems with ventilation,
- Direct Expansion (DX) or chilled water cooling coils, and
- Electric resistance, gas, or hot water heating coils.

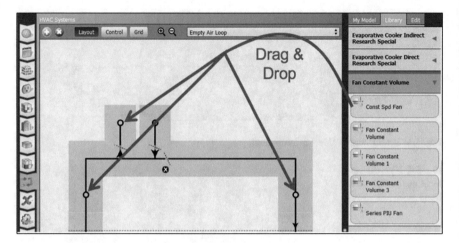

**Fig. 5.5**  Adding supply side Equipment in OpenStudio

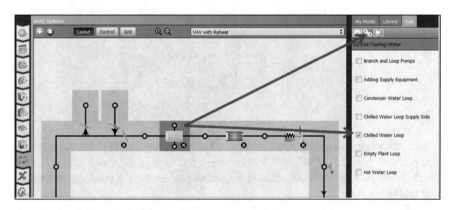

**Fig. 5.6**  Connecting a coil to a Plant Loop

To add a piece of Equipment to the supply side, first, select the [Library] Sub-Tab in the right-hand pane, then drag and drop Equipment onto a Node (Fig. 5.5). Not everything in the Library may be added to an Air Loop. OpenStudio attempts to help the user avoid nonsensical component placement by displaying a warning dialog and refusing to add the Object. This prevents the creation of invalid Air Loops that EnergyPlus cannot simulate.

Recall from Chap. 4 that Objects such as chilled water cooling coils and hot water heating coils must be connected to Plant Loops. For these components, Plant Loops supply the hot or cold water that the coils require for heat transfer. To connect a coil to a Plant Loop, first click on the coil Object on the diagram to select it. Next, click the [] Sub-Tab under the [Edit] Sub-Tab and check the box to select a Plant Loop. Once a coil is connected, a circle will appear above and below the coil. In Fig. 5.6, the Chilled Water Coil (blue) is connected to the Plant Loop named "Chilled Water

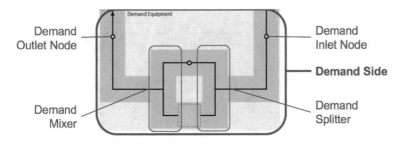

**Fig. 5.7**  Empty demand side Air Loop detail

Loop," while the Hot Water Coil (red) is not yet connected, as indicated by the absence of circles above and below the coil icon.

Creating Plant Loop linkages is just one of many required steps that the template systems from Chap. 4 performed for us automatically. However, understanding how such connections are made from Air Loop supply side components is important to understand when creating systems that may not be represented by an HVAC template.

### 5.2.1.2   Air Loop Demand Side

The demand side of an Air Loop is responsible for:

- Consuming conditioned air from the supply side via the demand inlet Node,
- Distributing it through one or more Thermal Zones, and
- Returning it to the supply side via the demand outlet Node.

As shown in Fig. 5.7, the demand side of an Air Loop includes a splitter and a mixer. The splitter distributes the incoming supply airflow through multiple branches. The mixer recombines the flow from the branches prior to returning it to the supply side for reconditioning.

Air from the Air Loop is supplied to Thermal Zones via Air Terminal Objects. An Air Terminal may be as simple as an air diffuser or more complex like a VAV terminal with integrated reheat. Air terminals are added to Air Loops like other components by dragging them from the ▬ Sub-Tab into the "Drag from Library" drop zone. Each Thermal Zone must be attached via an Air Terminal or the simulation will not run.

Recall from Chap. 4 that Thermal Zones may be added by dragging and dropping from Thermal Zone Objects from the ▬ Sub-Tab or by clicking on the demand splitter or mixer. The latter method allows the user to check boxes next to the Thermal Zones, quickly adding several at the same time (Fig. 5.8). As an added convenience, OpenStudio automatically pairs additional Thermal Zones with the same Air Terminal used with the first Thermal Zone. When adding multiple Thermal Zones, best practice is to attach the first Thermal Zone and Air Terminal pair, then check the boxes to attach the rest of the Zones.

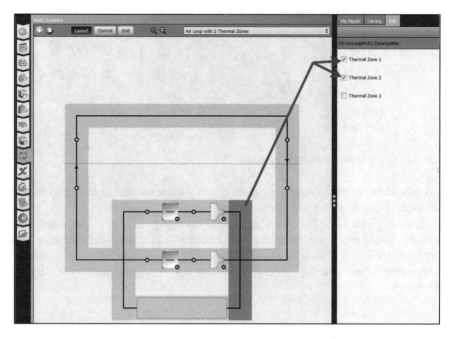

**Fig. 5.8** Connecting demand side Air Loop with two Thermal Zones and Simple Air Diffusers

## 5.2.2   Sizing Air Loops

Recall our high-level discussion of HVAC autosizing in Chap. 4. The sizing of Air Loops begins with the determination of heating and cooling loads for associated Thermal Zones and temperature setpoints for extreme days. One detail we glossed over in Sect. 4.5 was the specification of Thermal Zone sizing parameters. These are defined on the Thermal Zones (▦) Tab, as shown in Fig. 5.9 below and specify key supply air conditions for the Air Terminals serving each Thermal Zone. Knowing the Thermal Zone supply air conditions, EnergyPlus can determine the airflow rates required to achieve each Thermal Zone's setpoint for the prevailing thermal loads.

The supply side airflow that must be accommodated by an Air Loop is the sum of all individual Thermal Zone flow rates. This requisite airflow is considered along with the outdoor air (OA) ventilation requirements and the supply air sizing design targets. The sizing parameters associated with a particular Air Loop may be accessed by clicking on the dashed supply/demand line for the Loop shown in Fig. 5.10. These sizing parameters govern the Air Loop design supply air conditions, meaning the conditions delivered by the system before the Air Terminals, sometimes referred to as the deck temperature. For many types of systems, the deck temperature will be different than the Thermal Zone supply air temperature, because the Zone terminal units provide further heating or cooling in order to trim to each Thermal Zone's specific load.

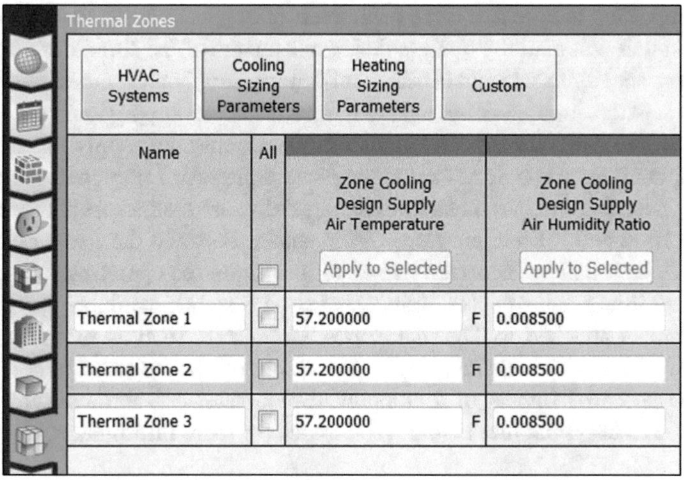

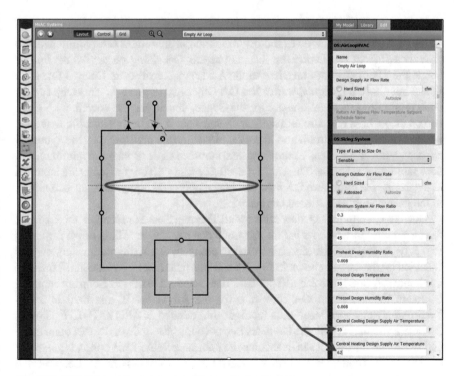

**Fig. 5.9**  Thermal Zone design supply air sizing parameters

**Fig. 5.10**  Specifying sizing parameters for a particular Air Loop

It is important to note that these parameters only govern how the system is sized, and not how it will actually operate during the course of the simulation. The sizing parameters specify the design heating and cooling supply air temperature, however there are separate input parameters that govern the actual supply temperature during operation. The input Objects and parameters governing the supply air temperature during operation will be described in the forthcoming Air Loop control discussion. For now, note that the parameters governing sizing are distinct from those controlling operation and if there are large discrepancies between the two sets of inputs, then the system will be incorrectly sized for the requested operating range.

Another important aspect of sizing Air Loops is setting available parameters for Air Terminals attached to Thermal Zones. Parameters such as minimum damper position of VAV terminals can impact the sizing of airflow rates. When in doubt about the potential impact of a component parameter on the autosizing process, refer to the EnergyPlus Input Output Reference for more information.

### 5.2.2.1   Outdoor Air (OA) Sizing

In addition to the determination of heating and cooling airflow requirements, the sizing of an Air Loop involves the determination of minimum OA requirements to provide sufficient ventilation for occupant health. OA sizing starts with the Spaces inside the Thermal Zones attached to the Air Loop. Recall from Chap. 3 that each Space may have a design specification OA Object assigned to it. This Object contains minimum OA requirements for the Space, which may be specified on a per person or per area basis, in terms of air changes per hour, or as a specific flow rate. At the beginning of a simulation, these requirements are multiplied by the appropriate occupancy level, area, volume, etc. and summed to determine the minimum OA flow rate for each Space. The values for all Spaces in a Thermal Zone are summed to determine the minimum OA flow rate that supply-side Equipment must condition to the sizing supply air design targets.

Once the minimum OA flow rates for all Thermal Zones are known, the sizing of the minimum OA flow rate for the Air Loop can happen. Scrolling down further in Fig. 5.10 reveals that the algorithm used to compute OA sizing parameters may be specified by the user as shown in Fig. 5.11. If this field is set to "ZoneSum," then the minimum OA flow rates for all Thermal Zones attached to the Air Loop are summed to create the minimum OA flow rate for the Air Loop. If this field is set to "VentilationRateProcedure," then an algorithm defined in ASHRAE 62.1[1] is used. Broadly speaking, this algorithm takes into account system diversity and other Zone and system characteristics to determine the minimum value. This option is generally used for multi-zone VAV Air Loops to ensure that all Zones on the Air Loop receive at least the minimum OA required, even during the worst-case conditions.

---

[1] ANSI/ASHRAE (2016).

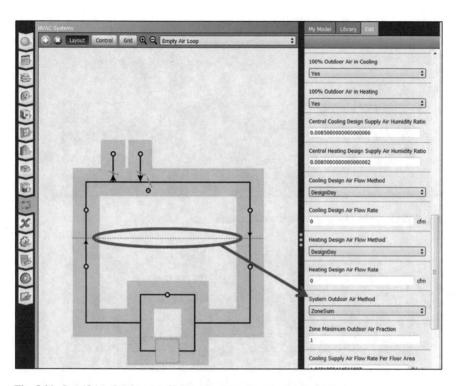

**Fig. 5.11** Specifying Outdoor Air Sizing Algorithm for a particular Air Loop

## 5.2.3 Air Loop Control

### 5.2.3.1 Air Loop Temperature Control

Air Loops are primarily controlled by setting the temperature (and sometimes humidity) of the supply air. This is accomplished using Setpoint Managers. These objects allow a user to define a temperature or humidity setpoint on a particular Node in the system. Some types of common Setpoint Managers are:

**Scheduled:** The setpoint follows a schedule, which may be set to a constant temperature. This method is most commonly used on VAV systems where the system is always supplying cool air to the terminals. In these systems, heating is performed by reheat coils integrated directly in the Air Terminals on a Zone-by-Zone basis.

**Warmest:** This approach increases the temperature of cool supply air until it just meets the cooling load of the Zone with the greatest cooling demand. This is also commonly used with VAV systems to model the "supply air temperature reset" control strategy. For these systems, the control method saves cooling energy while decreasing the use of reheat.

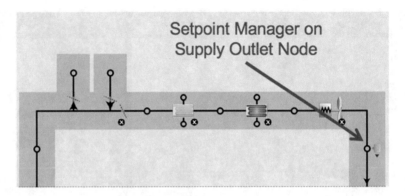

**Fig. 5.12**   Setpoint Manager on supply outlet Node

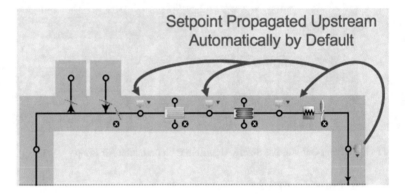

**Fig. 5.13**   Default OpenStudio Setpoint Manager propagation

**SingleZone Reheat:** This method adjusts the setpoint to whatever temperature is required to meet the Zone heating or cooling load. It is typically used on Air Loops that serve a single Thermal Zone and the Air Loop is designed to supply hot or cold air as necessary.

The most common configuration, shown in Fig. 5.12, places a single Setpoint Manager on the supply outlet Node. In this case, all of the Equipment upstream of the manager works to meet the setpoint.

Behind the scenes, EnergyPlus actually requires each component to have a setpoint established for its outlet Node. OpenStudio does this automatically by assigning the appropriate Setpoint Managers to all upstream Nodes, while keeping the diagram free of this additional, but necessary clutter. Figure 5.13 illustrates what OpenStudio is doing behind the scenes, and is shown here for completeness. The user is not required to add these additional Setpoint Managers in the user interface.

If a user does not want this default behavior, they may explicitly attach Setpoint Managers to the outlet Node of any supply side components. As an example, in Fig. 5.14 the user elected to set a cooling coil outlet temperature to 50 °F for

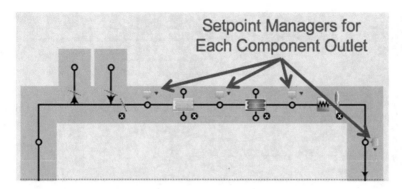

**Fig. 5.14** Adding Explicit Setpoint Managers for each supply side component

dehumidification purposes. A manager downstream of the fan could then set the heating coil and fan outlet to 60 °F to avoid condensation problems on uninsulated ductwork leading to the Air Terminals.

### 5.2.3.2 Air Loop Availability Control

So far, we have considered Setpoint Managers, which are primarily used to control conditions at Nodes throughout an Air Loop. OpenStudio also allows us to specify when control systems may operate. As shown in Fig. 5.15, a ▬▬ Button allows us to specify control characteristics for our HVAC systems.

The "HVAC Operation Schedule" field identifies a discrete (0/1) Schedule that determines when the selected Air Loop is allowed to operate. During times when this Schedule contains a one, the Air Loop operates normally. When the Schedule outputs a zero, Air Loop operation is determined by the "Use Night Cycle" field. This field may be used to specify one of three operating modes whenever the Schedule evaluates to zero:

**Follow the HVAC Operation Schedule:** The Loop will not operate.

**Cycle on Full System if Heating or Cooling Required:** The Loop only activates for 30 min when the Thermal Zone temperatures get too hot or too cold.

**Cycle on Zone Terminal Units if Heating or Cooling Required:** Any fan-powered Air Terminals attached to the Air Loop will run when the Thermal Zone temperatures get too hot or too cold, but the rest of the Air Loop will not.

### 5.2.3.3 Outdoor Air Control

Air Loops with an OA system contain three additional control considerations. The first is the Schedule when OA is supplied, and is made available when the OA Object in an Air Loop is selected as shown in Fig. 5.16. This is determined via the

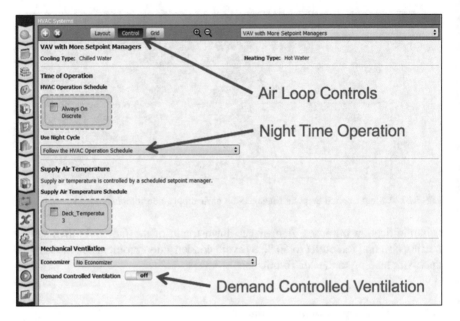

**Fig. 5.15** Specifying Air Loop availability control

discrete Schedule listed in the "Minimum Outdoor Air Schedule Name" field shown in the Figure. When this Schedule outputs a zero and the Air Loop is running, OA is not supplied. When this Schedule is set to one and the Air Loop is running, OA is supplied.

HVAC systems with gravity dampers generally have OA supplied whenever the system is operating. These systems should use an "Always On Discrete" schedule in this field. HVAC systems that have motorized dampers generally close the OA intake at night when the building is unoccupied. These systems may also use the "HVAC Operation Schedule" described above to turn off OA at night, even if the night cycle mode is set to manage heating or cooling loads.

The second OA control option relates to economizer operation. Economizers are devices that allow cool OA to be drawn into the system to provide free cooling when OA conditions are favorable. The type of economizer and associated settings may be found in the Controller OA Object, as shown in Fig. 5.16.

A third control option specifies whether the OA system uses demand-controlled ventilation (DCV). This is controlled via a switch on the controls page (Fig. 5.15). DCV is a control strategy that involves monitoring occupancy levels, typically via a proxy like $CO_2$, and reduces OA below the design flow rate when the occupancy levels are lower.

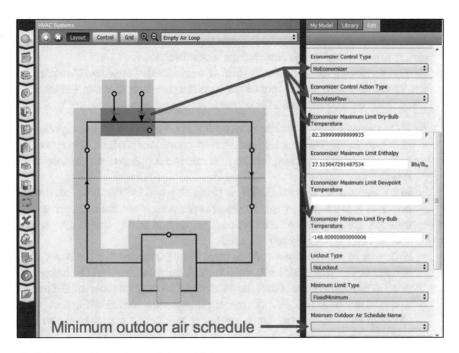

**Fig. 5.16** Specifying OA control for an Air Loop

### 5.2.4 Air Loop Simulation Process

The overall simulation process for each time step may be summarized as follows.

1. EnergyPlus calculates the heating or cooling load in the Thermal Zone.
2. The Air Terminals connected to the Thermal Zone convert this load to an airflow rate given the temperature of air set by the Zone sizing parameters.

   - This is critically important and deserves reiteration in bold text: **the airflow rate in an Air Loop is driven by the Air Terminals**. The parameters of these terminals are the first place to check when understanding flow rate behavior.

3. Once the flow rates for each terminal are determined, they are summed and the total request is passed to the Air Loop itself.

   - If Air Loop parameters limit the total flow rate below this amount, the Air Terminals will be supplied less than they request.

4. Two types of fans may be used in an Air Loop: constant volume or variable volume. The names are a bit misleading however, because both fans will modulate the airflow rate according to the volume of air requested by the Air Terminals. Remember the Air Terminals are in charge. The primary difference between the constant and variable volume fans is that the variable volume fan contains a fan

power curve that modifies the fan efficiency as the requested airflow moves away from the design point, while the constant volume fan simulates a constant efficiency regardless of the airflow rate moving through it.

5. After determining the total supply airflow rate, the Air Loop then uses parameters associated with the OA controller to determine how much of the total Air Loop flow rate will be OA and how much will be return air.

   - If the minimum OA schedule is zero for this time step, no OA is supplied.
   - If the Air Loop needs cooling and the OA conditions are favorable according to the economizer settings, the OA flow rate is calculated to get as close to the setpoint as possible.
   - In the event that no economizing is happening and DCV is enabled, the OA flow rate is calculated by multiplying the per person ventilation requirements for each Zone by the current occupancy level and adding this to the per area and air change per hour requirements.
   - Note that in a multi-Zone system, there is no guarantee that the appropriate amount of OA will reach each Air Terminal, only that the total OA being supplied to the system meets the total minimum OA requirements of all Thermal Zones at this time step. For this reason, it is prudent to review the time series OA outputs to ensure that Zones are being ventilated adequately at all times. Furthermore, mechanical design engineers and OpenStudio users often try to group similar Zones on the same system, to minimize the potential for over or under ventilation of a particular Zone on a multi-Zone system.
   - If DCV is not enabled, the OA flow rate is the design size OA flow rate for the Air Loop.

6. After the return air and OA flow rates are determined, the combined air temperature and humidity are determined by mixing these airstreams at their respective temperatures and humidity levels.
7. The mixed air is then passed to downstream components, which attempt to condition the air based on their outlet Node setpoints.
8. Hopefully, depending on the capacities of these components and their control settings, the supply outlet Node air achieves the desired design temperature and humidity values before it is passed to the Air Terminals.

Sometimes the components cannot hit the setpoints they have been assigned. Perhaps the components have been scheduled to be unavailable, or they simply don't have sufficient capacity. In these cases, the supply air may not be hot or cold enough to meet Zone thermal loads. In these cases, air in the Thermal Zones goes above or below the cooling or heating setpoint. Each hour that a Thermal Zone misses its heating or cooling setpoint beyond an allowable tolerance is defined as an unmet hour. Unmet hours are used as a quality metric to determine whether a Model is generally meeting its setpoints. As a rule of thumb, 350 or fewer total unmet heating or cooling hours during occupied times is considered the threshold for acceptable Model behavior.

**Fig. 5.17** Flow direction
in an empty Plant Loop

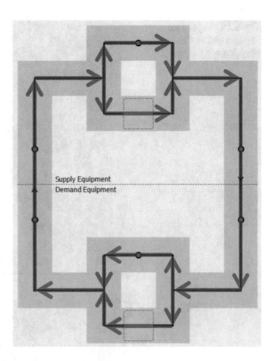

## 5.3   Plant Loops

A Plant Loop is a series of objects that represent a hydronic piping system inside of
a building. A loop may supply hot water, chilled water, or condenser water to various HVAC components within the building. These components turn this water into
a service like heating or cooling for some particular part of the building.

### 5.3.1   Plant Loop Configuration

In OpenStudio, a Plant Loop is represented conceptually by the diagram shown in
Fig. 5.17. Water always flows in the directions shown by the red arrows. A Plant
Loop is a closed system, so water circulates continuously without leaving.

As shown in Fig. 5.18, Plant Loops like Air Loops break down into supply and
the demand sides.

#### 5.3.1.1   Plant Loop Supply Side

The supply side of a Plant Loop is responsible for:

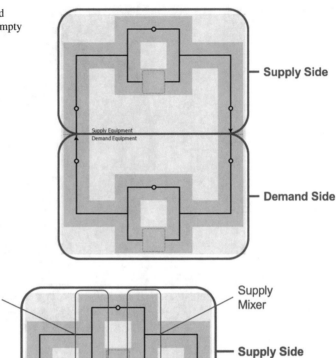

**Fig. 5.18** Supply and demand sides of an empty Plant Loop

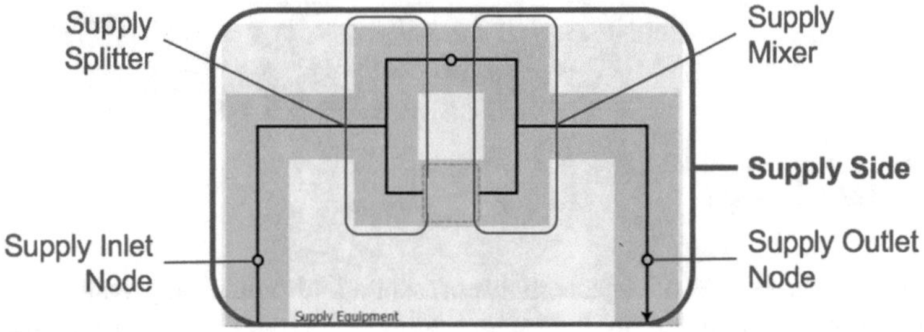

**Fig. 5.19** Empty Plant Loop supply side detail

- Consuming water returning from building's HVAC Equipment via the supply inlet Node,
- Heating or cooling it to the correct temperature, and
- Supplying it to the building's HVAC Equipment via a supply outlet Node.

An empty supply side portion of a Plant Loop is shown in Fig. 5.19. Supply side components may be added to Loop between the supply inlet and outlet Nodes.

The supply side of a Plant Loop contains a splitter and mixer. The splitter diverts the incoming flow through multiple branches. The mixer recombines the flow from the branches. For example, Fig. 5.20 illustrates a Plant Loop with two supply side chillers. A walkthrough of the process is as follows:

1. Water enters the supply side at the supply inlet Node and travels to the supply splitter,
2. The flow is divided between the two chillers based on the control sequence,
3. The chillers cool the water,
4. The supply mixer recombines the flows from the chillers, and
5. Water exits the supply side at the supply outlet Node.

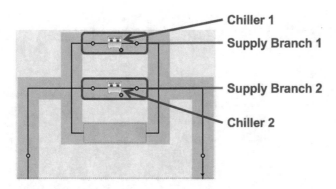

**Fig. 5.20** Plant Loop supply side containing two chillers

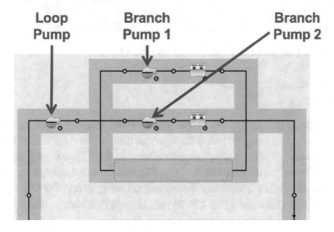

**Fig. 5.21** Plant Loop with one Loop and two branch pumps

In EnergyPlus, a Plant Loop must contain at least one pump on the supply side. A pump may be located either on the loop itself or on a branch as shown in Fig. 5.21. Just as we saw with Air Loop fans, pumps in EnergyPlus do not drive water flow around the loop. Instead, the water flow rate through the pumps is established by the flow requests made by the Equipment on the demand side of the Plant Loop. The pumps merely respond to the requested water flow rate and apply lookups for pressure drop and efficiency to account for the energy that would be consumed to move that much water. Loop pumps experience the full flow rate of the Plant Loop, but branch pumps only accommodate the flow rate through their particular branch. The Plant Loop simulation process is discussed further in Sect. 5.3.4, and is similar to the process used for Air Loop calculations.

In addition to pumps, common supply side Equipment may include:

- Chillers,
- Boilers,
- Cooling Towers,

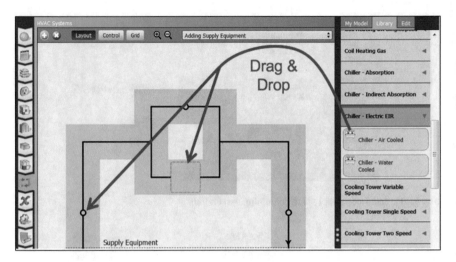

**Fig. 5.22** Adding Equipment to the supply side of a Plant Loop

- Fluid Coolers, and
- Heat Exchangers.

To add a piece of Equipment to the supply side, select the ▭ Sub-Tab in the right-hand pane, then drag and drop Equipment onto a Node (Fig. 5.22) or onto the "Drag from Library" drop zone. Not everything in the Library may be added to a Plant Loop. OpenStudio attempts to help the user avoid nonsensical component placement by displaying a warning dialog and refusing to add the Object. This prevents the creation of invalid Plant Loops that EnergyPlus cannot simulate.

### 5.3.1.2   Plant Loop Demand Side

The demand side of a Plant Loop is responsible for:

- Consuming conditioned water from the supply side via the demand inlet Node,
- Distributing it through the building's HVAC Equipment, and
- Returning it to the supply side via the demand outlet Node.

As shown in Fig. 5.23, the demand side of a Plant Loop includes a splitter and a mixer. The splitter distributes the incoming supply water flow through multiple branches. The mixer recombines the flow from the branches prior to returning it to the supply side for reconditioning.

Like the supply side of the Plant Loop, the demand side also includes a splitter and a mixer. The splitter distributes the incoming flow through multiple branches. The mixer recombines the flow from the branches before returning it to the demand outlet Node. Figure 5.24 below shows a Plant Loop with two Cooling Coils on the demand side.

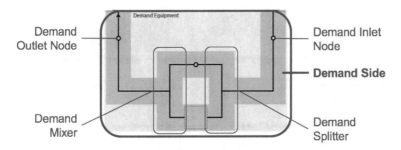

**Fig. 5.23**  Demand side details of an empty Plant Loop

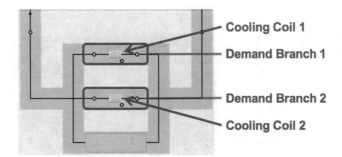

**Fig. 5.24**  Demand side of a Plant Loop serving two Cooling Coils

A walkthrough of this demand side is as follows:

1. Water enters the demand side at the demand inlet Node and travels to the demand splitter,
2. The flow is divided between the two cooling coils depending on their requested flow,
3. The cooling coils absorb heat from the building, warming the water,
4. The demand mixer recombines the flows from the cooling coils, and
5. Water exits the demand side at the demand outlet Node.

### 5.3.2   Plant Loop Sizing

The sizing of Plant Loops takes place after the sizing of Air Loops. This is because the heat transfer requirements of the Plant Loop working fluid are dictated by the heat exchange needs of the Air Loop. Figure 5.25 illustrates how air passes through the coil, exchanging heat with the working fluid provided by the Plant Loop.

Recall from previous sections that the Air Loop sizing process determines heating or cooling capacity requirements corresponding to the airflow rate and design supply temperature for each coil. Using the coil sizing parameters shown in Fig. 5.26

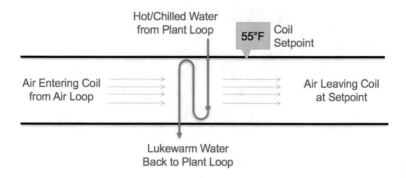

**Fig. 5.25** Air and Water Flow through a Coil

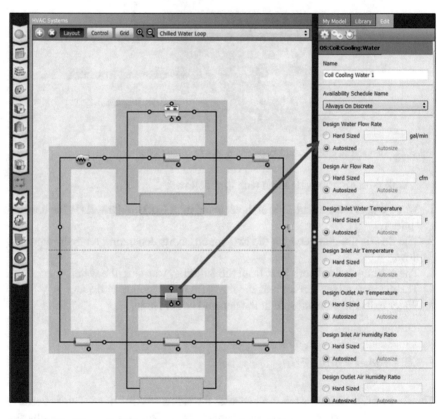

**Fig. 5.26** Plant Loop Coil Sizing parameters for hard or autosizing

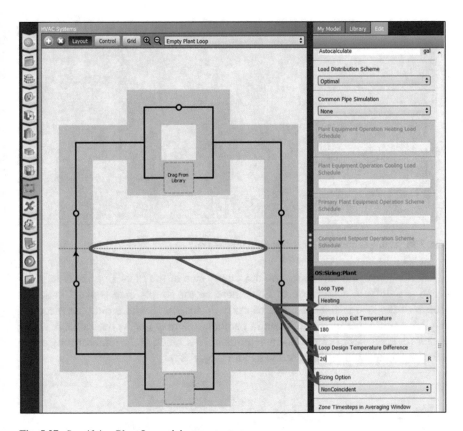

**Fig. 5.27** Specifying Plant Loop sizing parameters

and prescribed Plant Loop design temperatures, the water mass flow rate required to achieve a particular heating or cooling capacity may be calculated.[2]

Clicking the supply and demand dashed line on the diagram displays the Plant Loop parameters, including autosizing values. Figure 5.27 shows the Plant Loop sizing parameters, which include the design Loop supply temperature and desired temperature delta across the demand side. It is important to inspect these parameters for a given application, as they vary depending on the type of Loop - e.g. hot, cold, or condenser water. In addition, the previous discussion about Air Loop sizing applies here, in that the plant sizing parameters are distinct from the properties governing operation. It is important to coordinate sizing parameters with control parameters governing operation. Once the design flow rates and capacities are known for each Plant Loop coil, the flow rates are summed to determine the design flow rate and capacity for the Loop.

---

[2] Note in this particular example, all coil sizing parameters are set to autosize and will be computed using the process outlined previously. However, hard sized values could also have been selected for a specific piece of equipment.

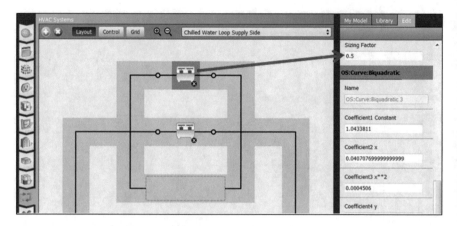

**Fig. 5.28** Sizing factor used for components installed on branches

At this point, the Equipment on the supply side of the Plant Loop may be auto-sized. By default, each piece of Equipment on the supply side is sized to accept 100% of the design flow rate and capacity. In instances where multiple pieces of Equipment are intended to operate in parallel, this may not be desirable. A sizing factor for each piece of Equipment determines how much of the total capacity parallel components will be designed to handle (Fig. 5.28).

### 5.3.3   Plant Loop Control

#### 5.3.3.1   Plant Loop Temperature Control

Plant Loops are primarily controlled by setting the temperature of the water at various points on the loop. As with Air Loops, this is accomplished using Setpoint Managers. These objects allow a user to define a temperature setpoint on a particular Node in the system. Some types of common Setpoint Managers used on Plant Loops include:

**Scheduled:** Setpoint follows a schedule, generally set at a constant temperature. This method is most commonly used on VAV systems, where the system is always providing cool air (e.g. 55°F) to the terminals, and heating, if needed, is provided by reheat coils in the Air Terminals on a Zone-by-Zone basis.

**Outdoor Air Reset:** Modifies the setpoint of the water by linearly interpolating between two points based on two corresponding OA conditions. This is used to apply the "OA temperature reset" control strategy, where for example chilled water temperature is increased when it is cold outside, because less cooling is likely to be needed, or hot water temperature is decreased when it is warm outside, because less heating is likely to be needed.

**Follow OA Temperature:** Sets the setpoint to the OA temperature (dry bulb or wet bulb) plus or minus a specified offset. Most commonly used to control the temperature leaving cooling towers.

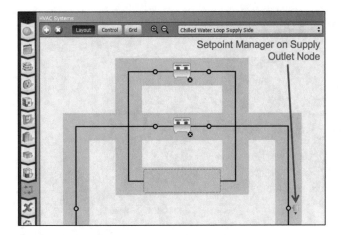

**Fig. 5.29** Plant Loop supply control

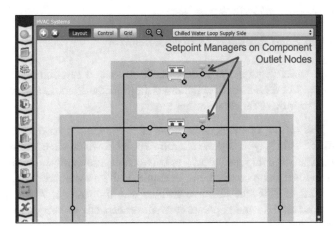

**Fig. 5.30** Multiple Setpoint Managers for component control

The placement of Setpoint Managers around the loop determines which type of control scheme is used. There are two main options: Loop level control and component setpoint control. The most common configuration, shown in Fig. 5.29, is to place a single Setpoint Manager on the supply outlet Node. In this case, if the loop needs to provide cooling to meet this setpoint, cooling Equipment on the loop will be dispatched. If the loop needs to provide heating to meet this setpoint, heating Equipment on the Loop will be dispatched. The order of dispatch will be described in the next Section.

The other type of control is component setpoint control. For this type of control, Setpoint Managers are attached to the outlet Nodes of each piece of heating or cooling Equipment (Fig. 5.30). Components will be dispatched to meet these outlet setpoints.

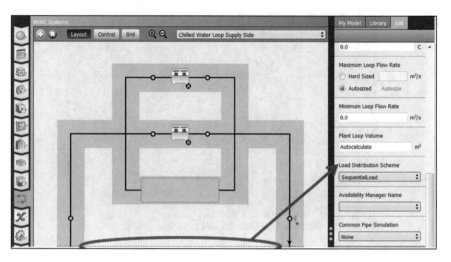

**Fig. 5.31** Plant Loop Load Distribution Scheme

### 5.3.3.2   Plant Loop Equipment Dispatch Control

The order in which Equipment is dispatched to meet a setpoint is based on the setting of the "Load Distribution Scheme" field in the Plant Loop, as shown in Fig. 5.31. Descriptions for each setting are:

**Optimal:** This setting operates each piece of Equipment at its optimal part load ratio (PLR). This is typically the operating point at which the equipment is most efficient. The last component operates between its minimum and maximum PLR in order to meet Loop demand. For example, if there were three chillers with an optimal PLR of 75%, the first two chillers would operate at 75% capacity and then the third chiller would operate at whatever capacity was required to meet the remaining demand.

**SequentialLoad:** With this approach, each piece of Equipment operates sequentially from top to bottom on the Plant Loop diagram. Each component is loaded to its maximum PLR. The last required component operates at a PLR between its minimum and maximum required to meet remaining Loop demand.

**UniformLoad:** This dispatch algorithm evenly distributes Loop demand across all available components.

**SequentialUniformPLR:** Using this method, components are loaded sequentially from top to bottom on the Plant Loop diagram. If the first unit cannot meet the required load, then a second unit is brought online and both unit PLRs are set equally. Additional units are added in this manner until the load is met.

**UniformPLR:** The final setting loads all Equipment to a uniform PLR. No Equipment is loaded below its minimum PLR, and one or more units may be shut off to keep the remaining units above minimum PLR.

### 5.3.3.3 Prescribing Plant Loop Load Ranges

Some dispatch methods limit certain pieces of Equipment between specified bands. For example, only operate chillers one and two when the load is less than 200 tons but operate chillers two and three when the load is between 200 and 500 tons. It is possible to define these load ranges and corresponding sequences of operation in EnergyPlus, but this feature is currently not implemented in the OpenStudio user interface.

## 5.3.4 Plant Loop Simulation Process

The simulation process for each Plant Loop is similar to how Air Loops are evaluated.

1. EnergyPlus calculates the flow rate demanded by each coil served by the Plant Loop.
2. These are summed to determine the total design flow rate that must be supplied by the Plant Loop for the next time step.
3. Two types of pumps that can be used on a Plant Loop: constant and variable speed. Constant speed pumps process 100% of the design flow rate.
4. If the coils are requesting the full design flow rate, then the entire flow passes through the coils.

   - When less coil flow is required, the difference is bypassed directly to the supply inlet Node. This bypass takes place via an implicit branch not shown in the diagram and behaves just as if a bypass pipe Object was dropped into a branch.

5. Based upon the selected dispatch scheme, the Plant Loop flow is split between each heating or cooling component accordingly.
6. Conditioned water flows are combined by the mixer before being returned to the supply outlet Node.

Sometimes components may not achieve their designated setpoints. This may occur due to component schedules, availability, or lack of capacity. In these cases, the water may not be hot or cold enough to provide sufficient coil heat transfer. These coils, in turn, may not provide sufficient conditioning for Zone loads, resulting in unmet hours. When troubleshooting unmet hours issues, it is often valuable to examine the temperatures and flow rates on Nodes throughout the Plant Loop to understand whether the various setpoints are being met.

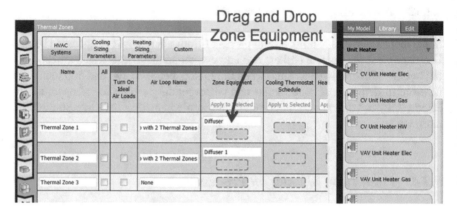

**Fig. 5.32** Adding Zone Equipment to a Thermal Zone

## 5.4  Zone Equipment

Zone Equipment is the third category of HVAC Equipment in EnergyPlus. As discussed in Chap. 4, each piece of Zone Equipment is dedicated to a single Thermal Zone, and operates based on its heating, cooling, and ventilation needs.

### 5.4.1  Zone Equipment Configuration

Recall that Zone Equipment is added to a Thermal Zone on the Thermal Zones (▣) Tab. It is dragged from the Library in the right-hand pane into a specific Thermal Zone's Equipment column (Fig. 5.32). Multiple pieces of Zone Equipment may be assigned to a single Thermal Zone.

A piece of Zone Equipment may be a single Object or contain child components such as fans, heating coils, and cooling coils. The objects that comprise a piece of Zone Equipment may be reviewed by clicking on the Object and scrolling through its properties as shown in Fig. 5.33.

### 5.4.2  Zone Equipment Sizing

As with Air Loops, the sizing of Zone Equipment begins with determination of heating and cooling loads and required airflow rates for the Thermal Zone it serves. The design flow rate is used to calculate the capacities and flow rates required for any heating, cooling, or ventilation components included within the Zone Equipment. Some types of Zone Equipment contain their own sizing parameters, so it is prudent to review these settings and ensure that they align with the Thermal Zone sizing parameters. Mismatched sizing design parameters may result in over- or under-sized Equipment.

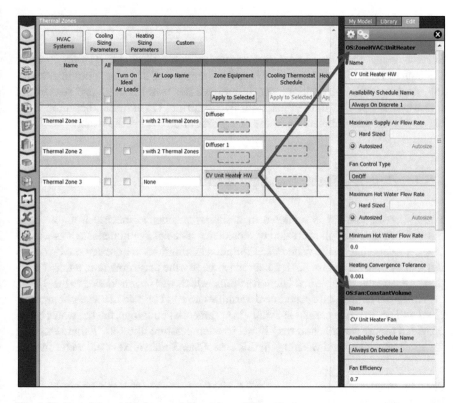

**Fig. 5.33** Examining child objects within a piece of Zone Equipment

## 5.4.3 Zone Equipment Control

Controls are specific to each type of Zone Equipment and are defined in the component
[Edit] Sub-Tab. Generally, Zone Equipment responds to the heating and/or cooling setpoints in the Thermal Zone being served. Some types of Zone Equipment cycle on and
off only to meet loads, while other types meet load and provide OA. Zone Equipment
that includes a fan may use a fan operation schedule to denote whether the fan cycles
with heating and cooling or runs continuously.

## 5.4.4 Zone Equipment Simulation Process

The simulation process for Zone Equipment is similar to that used for Air or Plant
Loops with the exception that the Equipment is sized to meet the heating or cooling
load of its associated Zone. In the event that multiple pieces of Zone Equipment are
attached to a single Thermal Zone, they are sized and simulated in order from top to
bottom. This means that if additional load remains after the first piece of Zone
Equipment is simulated, that load is passed along to the next piece of Equipment
listed in the Equipment column.

## 5.5   Service Water Systems

Service Water Systems, sometimes known as domestic water systems, are the systems in buildings that supply water for purposes of drinking, cooking, washing, flushing toilets, irrigation, etc. Sometimes this water is cold, and sometimes this water is heated. EnergyPlus, and, by extension, OpenStudio, can model the usage of cold and hot water, and the energy required to heat the water.

### 5.5.1   Defining Water Use Equipment

Water Use Equipment is modeled in a similar manner to electric Equipment or lights. First, a definition is created to represent a piece of Equipment such as a toilet or sink. Like other loads, Water Use Equipment definitions are created on the Loads (◎) Tab. As shown in Fig. 5.34, a user may specify the peak flow rate, a target temperature, and the sensible and latent fractions, which will be introduced to the Space.

The target temperature schedule determines how hot or cold the water being supplied will be. In many cases, especially in commercial buildings, the hot water being provided by the water heater may be set to a temperature like 140 °F that is too hot for some uses like hand washing. In this case, OpenStudio mixes cold water in with

**Fig. 5.34** Creating a Water
Use Equipment definition

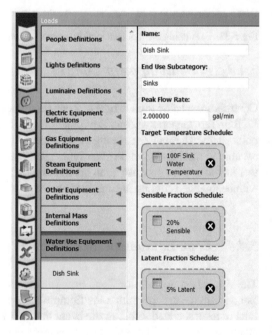

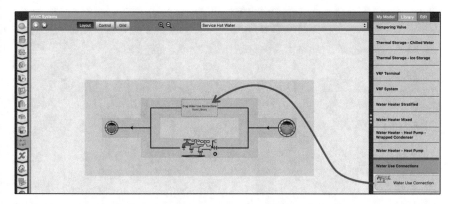

**Fig. 5.35** Adding a water use connection to the service Hot Water Loop

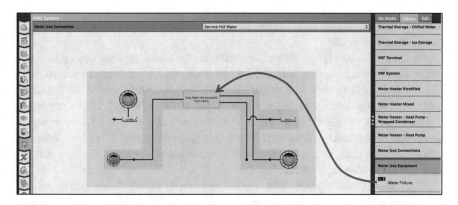

**Fig. 5.36** Adding a sink to an empty water use connection

the hot water to match the target temperature. For Equipment that uses cold water like toilets and irrigation, the target temperature schedule may be set to a value below that of the temperature supplied by the water mains.

## 5.5.2 Adding Water Use Equipment to a Model

Adding Water Use Equipment to the Model requires multiple steps on the HVAC (🔲 ) Tab. Begin by selecting "Service Hot Water" from the system selector at the top of the window. Water use Connection Objects are added to the Model by dragging them from the ▭ as shown in Fig. 5.35.

This Object connects the Water Use Equipment inside it with the cold-water main entering the building and, optionally, a hot water Loop. By clicking on the Water Use Connections Object, a new view appears (Fig. 5.36) into which Water

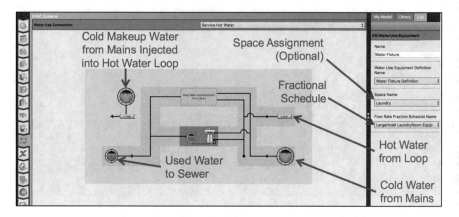

**Fig. 5.37**  Adding Equipment to a water use connection

Use Equipment may be dragged and dropped. Water use Equipment may be added from either ▢ or ▢.

Figure 5.37 illustrates a Water Use definition for a sink that has been added to a Water Use Connection diagram. Cold water is always supplied to every Water Use Connection. This is unheated water straight from the water mains. A fractional scheduled is required for each piece of Equipment that is added. Like other loads, the value of this schedule is multiplied by the peak flow rate at each time step to determine how much water is used. Water Use Equipment may be optionally added to a Space to reflect additional sensible and latent loads. If only cold water is being used, this step may be omitted.

For Water Use Equipment that requires hot water, an additional step is needed. The Water Use Connection containing the Water Use Equipment must be attached to the demand side of a Plant Loop with a water heater or boiler. Figure 5.38 shows a service water Loop with a hot water heater on the supply side and a Water Use Connection on the demand side. EnergyPlus requires a pump on any Plant Loop.

> **Tip**: Some buildings do not contain a recirculation pump, instead relying on mains pressure to drive flow. To model these cases, a variable speed pump can be added to the Plant Loop with a pressure rise of zero so that the pump does not consume power.

Because water heaters in OpenStudio cannot be auto-sized, capacities must be entered manually. This means that a water heater may not be able to keep up with the demand from the Water Use Connections. For unmet heating and cooling hours, there is a convenient internal variable in EnergyPlus that tracks whether setpoints

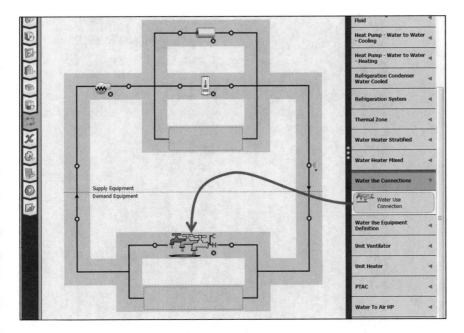

**Fig. 5.38**   Adding a hot water connection to a Hot Water Loop

are met. There is no equivalent for whether or not the setpoint of Water Use Equipment is met. For this reason, it is advisable to review the detailed time series output of the leaving temperature of Water Use Equipment to ensure that it is meeting the target temperature schedule. If it is not, the size of the water heater may need to be increased.

## 5.6   Checkpoint Six: Air Loops

In this and subsequent Chap. 5 exercises we are going to build up from scratch the VAV systems that were created using templates in Chap. 4. This will allow us to gain familiarity with OpenStudio's capability for modeling complex HVAC systems, while comparing our results from the known results we achieved in the previous exercise. If you followed our instructions from Chap. 4, you made a copy of your rezoned school Model called *MyPrimarySchoolZoned.osm*. Open this Model and proceed to the HVAC (■) Tab to begin. Alternately, you can open *MyPrimarySchoolHVAC.osm* and delete all Air and Plant Loops from that Model with the ■ Button.

We need to begin by creating an empty Air Loop. Empty loops are found amidst the other template systems. Use the ■ Button on the HVAC (■) Tab to add one as shown in Fig. 5.39.

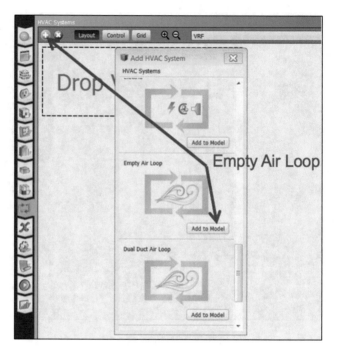

**Fig. 5.39** Adding an empty Air Loop

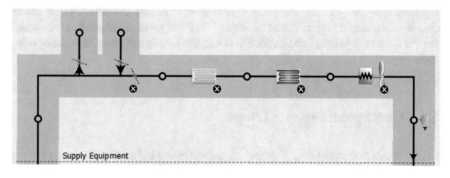

**Fig. 5.40** Air Loop with initial supply Equipment components

Once the empty Loop has been added follow these steps:

1. Add Objects to the Loop by dragging them from the ▢ onto Nodes as shown in Fig. 5.40:

   (a) OA
   (b) Coil Cooling Water
   (c) Coil Heating Water
   (d) Fan Variable Volume

2. Click on the dashed supply/demand line in the Loop to open the Loop Object.
3. Name it "My VAV Air Loop."

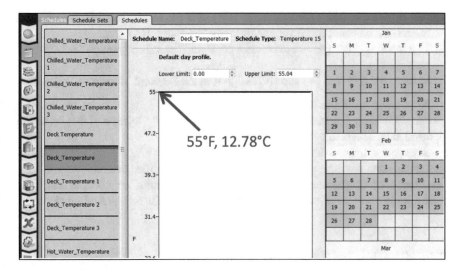

**Fig. 5.41**  VAV outlet setpoint temperature

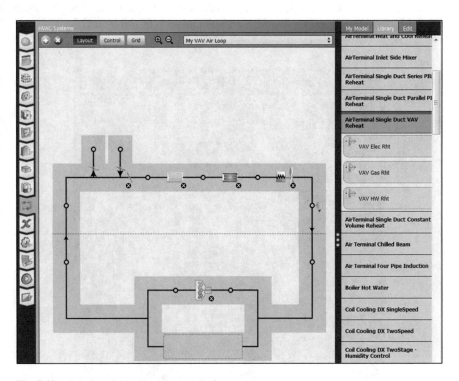

**Fig. 5.42**  Air Loop with VAV reheat terminal

1. Add a "Setpoint Manager Scheduled" Object to the supply outlet Node. We need to assign a constant 55 °F supply air temperature to this Object.
2. Use the Schedules (∎) Tab to create a constant "Deck Temperature" Schedule as shown in Fig. 5.41.
3. Return to the HVAC (∎) Tab and assign it to the Setpoint Manager's Schedule.

Next navigate in the ▬▬ to locate the "AirTerminal Single Duct VAV Reheat" terminal Objects. Select the "Hot Water (HW) Reheat" Object and drag it to the demand side. The result should now look like Fig. 5.42.

Save your work, and then compare this diagram with the VAV air handler from Chap. 4. Can you spot anything missing? Full marks if you identified that the Thermal Zone and Plant Loop Connection Nodes are missing from both coils. We will correct all three omissions in the next exercise, but we have successfully built our first Air Loop by hand.

## 5.7  Checkpoint Seven: Plant Loops

In this exercise we will add chilled, condenser, and hot water Loops to our Model by building them up from individual Objects. Begin by adding a chilled water Loop. Use the ∎ Button on the HVAC (∎) Tab to create an empty Plant

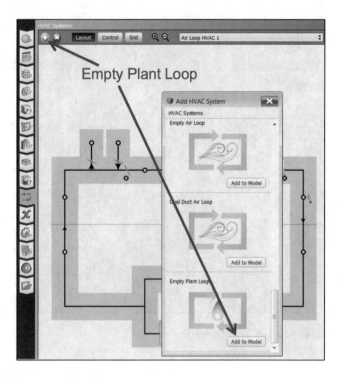

**Fig. 5.43** Add an empty Plant Loop

Loop, as shown in Fig. 5.43. Click on the dashed supply and demand line to open the Plant Loop Object in the editor and change its name to "My Chilled Water Loop."

1. Add Objects to your Loop by dragging them from the [Library] onto Nodes as shown in Fig. 5.44:

    (a)  Pump Variable Speed
    (b)  Water Cooled Chiller Electric EIR
    (c)  Bypass Pipe Adiabatic
    (d)  Setpoint Manager Scheduled

2. Name the chiller "My Water Cooled Chiller."
3. Add a Setpoint Manager that will maintain a constant supply water temperature of 44 °F.
4. Click on the supply/demand line to set the Loop sizing parameters, ensuring that:

    (a)  The Loop type is Cooling

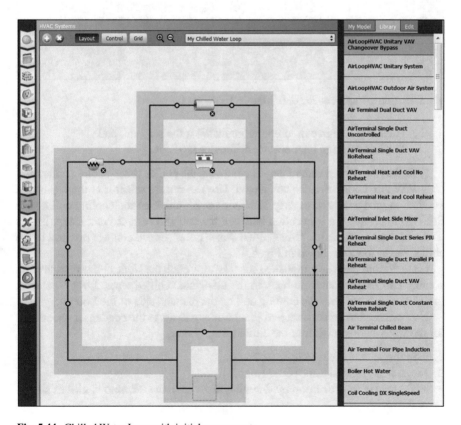

**Fig. 5.44**  Chilled Water Loop with initial components

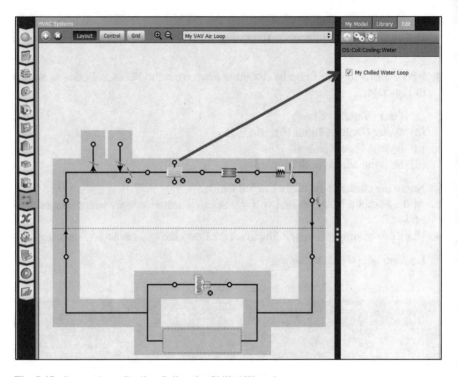

**Fig. 5.45** Connecting a Cooling Coil to the Chilled Water Loop

(b) The Loop design exit temperature matches the setpoint, and
(c) The Loop design temperature difference is 12 °R.

Now that the chilled water Loop is complete, the chiller water-cooling coils from "My VAV Air Loop" may be connected. Use the system selector at the top of the window to navigate back to "My VAV Air Loop." Click on the cooling coil and ▦ Sub-Tab to establish a connection between the coil and correct Plant Loop. In this case, we want to check the "My Chilled Water Loop" box to connect this coil to the chilled water loop, as shown in Fig. 5.45.

Once this has been accomplished successfully, click the Node above or below the cooling coil icon to navigate back to the associated chilled water Loop. It should look like Fig. 5.46, with the cooling coil on the demand side of the Loop.

Now we need to create a second Plant Loop to serve as the condenser water loop for the water-cooled chiller in the chilled water Loop.

1. Add the second Plant Loop
2. Name it "My Condenser Water Loop."
3. Add Objects to your Loop by dragging them from the ▦ onto Nodes as shown in Fig. 5.47:

(a) Pump Constant Speed

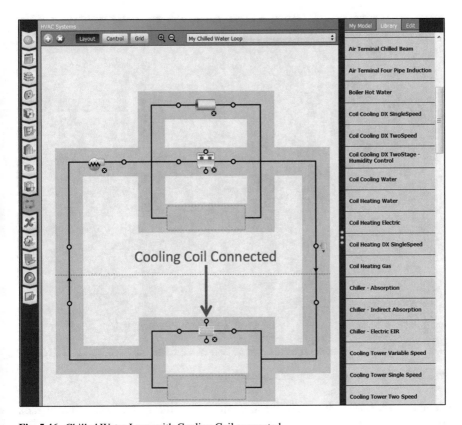

**Fig. 5.46** Chilled Water Loop with Cooling Coil connected

    (b) Cooling Tower Single Speed

    (c) Bypass Pipe Adiabatic

    (d) Setpoint Manager Follow Outdoor Air Temperature

4. Ensure that the setpoint:

    (a) Follows the Outdoor Air Wet Bulb with an offset of 4 °R,

    (b) Has a minimum of 70 °F, and

    (c) Has a maximum of 95 °F.

5. Click on the supply/demand line to set the Loop sizing parameters, ensuring that:

    (a) The Loop type is Condenser

    (b) The Loop design exit temperature to 85 °F, and

    (c) The Loop design temperature difference is 10 °R.

Once the condenser Loop has been completed, attach it to the chiller called "My Water-Cooled Chiller" from "My Chilled Water Loop" to the demand side of the Loop. Click on ▬▬ and drag "My Water-Cooled Chiller" to the

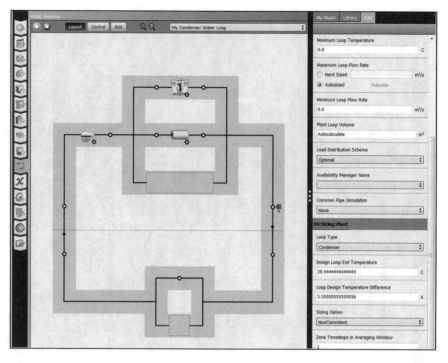

**Fig. 5.47** Condenser Water Loop with initial components

demand side. After successfully attaching the chiller, the Loop should look like Fig. 5.48.

> **Tip:** If you do not see the Node links above and below the chiller, you likely selected a second chiller from the ▭ and have accidentally added it to the model.

We still need to add a Loop to serve hot water to our hot water coil.

1. Add the third Plant Loop
2. Name it "My Hot Water Loop."
3. Add Objects to your Loop by dragging them from the ▭ onto Nodes as shown in Fig. 5.49:

   (a) Pump Variable Speed
   (b) Boiler Hot Water
   (c) Bypass Pipe Adiabatic

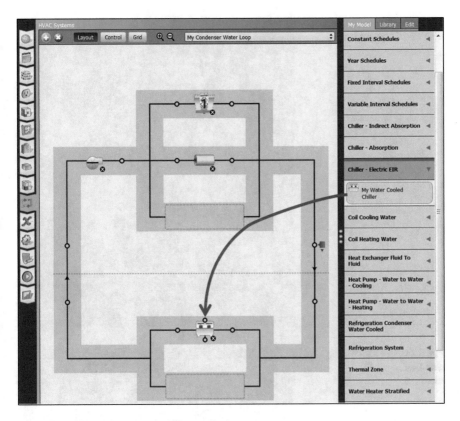

**Fig. 5.48**  Condenser Loop with chiller attached

(d) Setpoint Manager Scheduled

4. Ensure that the Setpoint Manager Schedule is 140 °F at all times.
5. Click on the supply/demand line to set the Loop sizing parameters, ensuring that:

   (a) The Loop type is Heating
   (b) The Loop design exit temperature matches the setpoint, and
   (c) The Loop design temperature difference is 20 °R.

Now that the hot water loop is ready, navigate back to "My VAV Air Loop" and follow the same process used to connect the cooling coil to connect the heating coil and air terminal to the hot water Loop. Verify that both the main heating coil and the coil inside the reheat terminal are connected. Afterward, the hot water Loop should look like Fig. 5.50.

Now that the reheat coil in the reheat terminal has been connected to the hot water loop, Thermal Zones can be assigned to the Air Loop. Navigate to "My VAV

**Fig. 5.49** Hot Water Loop
with initial components

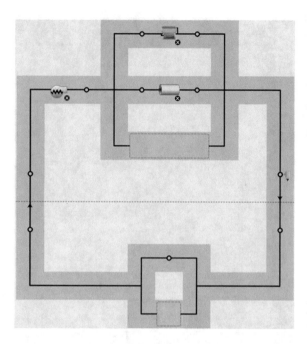

**Fig. 5.50** Hot Water Loop
with Heating Coils
connected

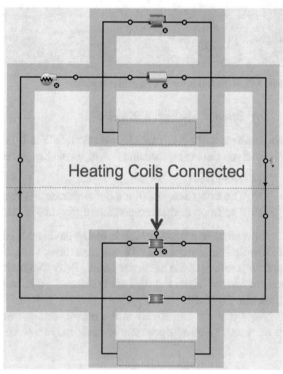

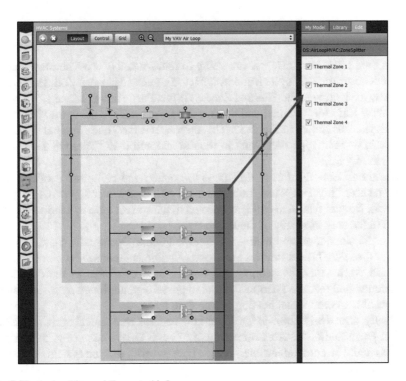

**Fig. 5.51**   Assign Thermal Zones to Air Loop

Air Loop." In the Air Loop, click on the demand splitter and use the check boxes to assign all of the Thermal Zones to the Air Loop, as shown in Fig. 5.51.

**Tip**: The reason we did not add Thermal Zones immediately after initial creation of the Air Loop is that the heating coils in the VAV terminals had not yet been connected to the hot water Loop. Adding additional Zones at that point would have automatically brought in the same, unconnected, terminal units. This would require us to manually assign each terminal coil to the hot water Loop.

Assuming this process was performed correctly, the hot water Loop should now have five heating coils attached to the demand side – one main coil for the VAV and four terminal reheat coils.

## 5.8   Checkpoint Eight: Zone Equipment

In this final HVAC exercise, we are going to remove one of the Thermal Zones in the Model from the VAV system and assign it to a dedicated piece of Zone Equipment. Begin by navigating to the Thermal Zones (▣) Tab. For "Thermal Zone 1," select the "VAV HW Rht" Air Terminal and remove this from the Zone using the ⊠ Button in the edit pane, as shown in Fig. 5.52. This disconnects the Zone from our Air Loop.

Locate a "Four Pipe Fan Coil" in the 🔳 and add it to "Thermal Zone 1," as shown in Fig. 5.53.

Use the 🔳 Sub-Tab in the Edit pane to connect the coils inside of the Zone Equipment to "My Hot Water Loop" and "My Chilled Water Loop" as shown in Fig. 5.54. Be sure that each coil is connected to the correct loop by reading the type of coil in the dark grey area of the Edit pane.

Run the simulation and examine the now familiar OpenStudio Results report. Look at the HVAC Load Profiles and Zone Conditions reports shown in Figs. 5.55 and 5.56. Verify that the heating and cooling load correlates to the outside air temperature and that the Thermal Zones are properly conditioned by reviewing the number of hours for unmet heating and cooling loads.

Finally, armed with knowledge about how detailed mechanical systems are created in OpenStudio, the Air Loops (Fig. 5.57) and Plant Loops (Fig. 5.58) detail reports should be comprehensible. Review both to verify that the systems were configured as intended.

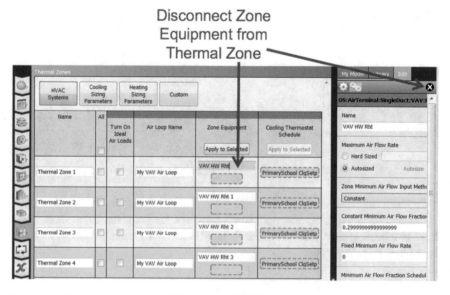

**Fig. 5.52** Disconnecingt the Thermal Zone from the Air Loop

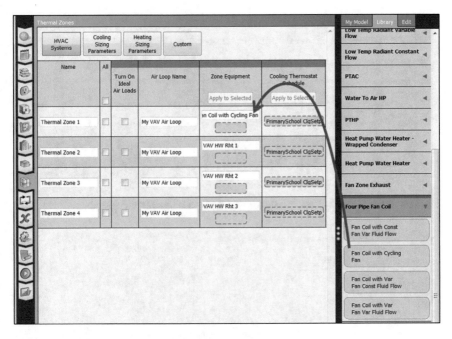

**Fig. 5.53** Adding Four Pipe Fan Coil Zone Equipment

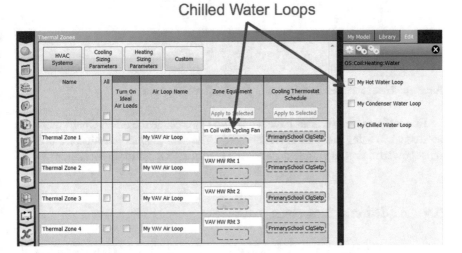

**Fig. 5.54** Connecting Zone Equipment Coil to a Plant Loop

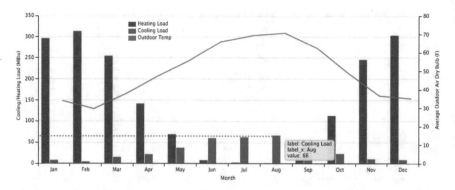

**Fig. 5.55**  HVAC Monthly Load Profile for school with advanced HVAC systems

| Zone | Unmet Htg (hr) | Unmet Htg - Occ (hr) | < 56 (F) | 56-61 (F) | 61-66 (F) | 66-68 (F) | 68-70 (F) | 70-72 (F) | 72-74 (F) | 74-76 (F) | 76-78 (F) | 78-83 (F) | 83-88 (F) | >= 88 (F) | Unmet Clg (hr) | Unmet Clg - Occ (hr) | Mean Temp (F) |
|---|---|---|---|---|---|---|---|---|---|---|---|---|---|---|---|---|---|
| THERMAL ZONE 1 | 27 | 0 | 0 | 196 | 851 | 416 | 1087 | 569 | 647 | 4272 | 369 | 353 | 0 | 0 | 0 | 0 | 72.5 (F) |
| THERMAL ZONE 2 | 56 | 0 | 0 | 538 | 1350 | 457 | 2161 | 949 | 792 | 2483 | 20 | 0 | 0 | 0 | 0 | 0 | 70.0 (F) |
| THERMAL ZONE 3 | 57 | 0 | 0 | 548 | 1349 | 465 | 2149 | 939 | 795 | 2495 | 20 | 0 | 0 | 0 | 0 | 0 | 70.0 (F) |
| THERMAL ZONE 4 | 106 | 1 | 0 | 199 | 1512 | 453 | 2087 | 992 | 917 | 2592 | 8 | 0 | 0 | 0 | 0 | 0 | 70.3 (F) |

| Zone | < 30 (%) | 30-35 (%) | 35-40 (%) | 40-45 (%) | 45-50 (%) | 50-55 (%) | 55-60 (%) | 60-65 (%) | 65-70 (%) | 70-75 (%) | 75-80 (%) | >= 80 (%) | Mean Relative Humidity (%) |
|---|---|---|---|---|---|---|---|---|---|---|---|---|---|
| THERMAL ZONE 1 | 4383 | 1389 | 1507 | 978 | 417 | 62 | 24 | 0 | 0 | 0 | 0 | 0 | 29.5 (%) |
| THERMAL ZONE 2 | 5327 | 822 | 761 | 640 | 605 | 345 | 205 | 55 | 0 | 0 | 0 | 0 | 28.2 (%) |
| THERMAL ZONE 3 | 5328 | 818 | 763 | 639 | 606 | 346 | 204 | 56 | 0 | 0 | 0 | 0 | 28.2 (%) |

**Fig. 5.56**  Zone conditions for school with advanced HVAC systems

Hopefully these exercises have helped you appreciate the effort OpenStudio can save us when using template HVAC systems. The next Chapter introduces one of the most powerful time saving features of the platform - OpenStudio Measures.

## 5.9   Additional Exercises

Recommended additional exercises involving the Checkpoint Seven Model include further study of HVAC system control and sizing.

- **Be sure to keep a "clean" copy of the Checkpoint Seven model before proceeding with these activities.**
- Change the supply air temperature setpoint for the VAV system.

My VAV Air Loop

| Object | Sizing | Sizing Units | Description | Value | Value Units | Count |
|---|---|---|---|---|---|---|
| *(supply)* | | | | | | |
| OS:AirLoopHVAC:OutdoorAirSystem | Autosized | cfm | Minimum Outdoor Air Flow Rate | Autosized | cfm | |
| OS:Coil:Cooling:Water | Autosized | gal/min | Plant Loop | My Chilled Water Loop | | |
| OS:Coil:Heating:Water | Autosized | gal/min | Plant Loop | My Hot Water Loop | | |
| OS:Fan:VariableVolume | Autosized | cfm | Pressure Rise | 4.09 | inH_{2}O | |
| OS:SetpointManager:Scheduled | | | Control Variable - Temperature | 55.0 to 55.0 | F | |
| *(demand)* | | | | | | |
| Thermal Zones | | | Total Floor Area | 55,197 | ft^2 | 3 |
| Thermal Zones | | | thermostat ranges for cooling | 75.2 to 80.6 | F | |
| Thermal Zones | | | thermostat ranges for heating | 60.8 to 69.8 | F | |
| Terminal Types Used | | | OS:AirTerminal:SingleDuct:VAV:Reheat | | | 3 |
| *(controls)* | | | | | | |
| HVAC Operation Schedule | | | | Always On Discrete | | |
| Night Cycle Setting | | | | StayOff | Choice | |
| Economizer Setting | | | | NoEconomizer | Choice | |
| Demand Controlled Ventilation Status | | | | false | Bool | |

**Fig. 5.57** Air Loops detail for school with advanced HVAC systems

My Chilled Water Loop

| Object | Sizing | Sizing Units | Description | Value | Value Units | Count |
|---|---|---|---|---|---|---|
| *(supply)* | | | | | | |
| OS:Pump:VariableSpeed | Autosized | gal/min | Rated Power Consumption | Autosized | W | |
| OS:SetpointManager:Scheduled | | | Control Variable - Temperature | 44.1 to 44.1 | F | |
| OS:Chiller:Electric:EIR | Autosized | Btu/h | Reference COP | 5.50 | | |
| OS:Chiller:Electric:EIR (cont) | | | Chiller Source | My Condenser Water Loop | | |
| *(demand)* | | | | | | |
| OS:Coil:Cooling:Water | | | Air Loop | My VAV Air Loop | | |
| Air Terminal Connections | | | | | | 1 |
| *(controls)* | | | | | | |
| Loop Flow Rate Range | Autosized | gal/min | Minimum Loop Flow Rate | 0.0 | gal/min | |
| Loop Temperature Range | | | | 32.0 to 212.0 | F | |
| Design Loop Exit Temperature | | | | 44.00 | F | |
| Loop Design Temperature Difference | | | | 12.00 | R | |

**Fig. 5.58** Plant Loops detail for school with advanced HVAC systems

- Change the deck temperature schedule from 55 °F to 60 °F.
- Compare the heating and cooling energy with Checkpoint Seven. Can you explain the reason for the changes?
- Explore different equipment staging methods.
  - Add a second Chiller to the Chilled Water Loop.
  - Change the sizing factor for each Chiller to 0.5.
  - Change the Load Distribution Scheme to UniformPLR for the Chilled Water Loop.
  - Run the simulation and look at the time series results for each Chiller.
    - Can you tell that the load is split between both Chillers?
  - Change the Load Distribution Scheme to SequentialLoad for the Chilled Water Loop.
  - Run the simulation and look at the time series results for each Chiller.
    - Is one of the Chillers loaded first?
    - Can you tell which Chiller runs for more hours in the year?
- Study different Fan control approaches.
  - Replace the Variable Volume Fan from the VAV system with a Constant Volume Fan.
  - Compare the fan energy with Checkpoint Seven. Can you explain the reason for the change?
- Experiment with the impact of design targets on system sizing.
  - Change the Central Heating Supply Air Temperature and Central Cooling Supply Air Temperature in the VAV system to 65 °F.
  - Leave the Deck Temperature set to 55 °F.
  - Run a simulation and note the number of unmet hours and system sizes.
  - Now change the Load Distribution Scheme to UniformPLR for the chilled water loop.
  - Run the simulation again.
    - Was there an increase or decrease in unmet hours? If so, why?
    - How did the size of the chiller change? If so, how did changing the sizing of the air loop affect the size of components on a chilled water loop?

# Reference

ANSI/ASHRAE Standard 62.2-2016 ventilation for acceptable indoor air quality, ASHRAE, 2016

# Chapter 6
# OpenStudio Measures

## 6.1 Introduction to OpenStudio Measures

Chapter 1 briefly mentioned the concept of OpenStudio Measures. These small scripts written in the Ruby[1] programming language are a unique feature of OpenStudio and are key to the platform's extensibility. Measures are most often used to implement model transformations that correspond to Energy Efficiency (EE) measures – hence their name. However, as we shall see, they can be used to query and transform OpenStudio models and associated data in a variety of ways. Throughout this text, whenever you see the word Measure capitalized, know that we're referring to an OpenStudio script as opposed to an EE measure.

Like their namesake, Measures are frequently used to apply an EE technology to a building Model in a simple, self-contained operation. For example, if a designer wishes to assess the potential for Energy Recovery Ventilation (ERV) to save energy in a building, they can select the ERV Measure and apply it to their Model. Figure 6.1 illustrates and HVAC system before and after the Measure is applied. Other examples include modification of an HVAC system's coefficient of performance (COP), alterations to window-to-wall ratio (WWR), and alteration of insulation thermal resistance (R-Values).

Simple parametric substitutions are generally within the capability of other modeling tools and scripting languages. Since OpenStudio Measures are written in the full-featured Ruby scripting language, and have access to components in OpenStudio's Object Model, Measures can do much more. One of the most significant examples is shown in Fig. 6.2.

---

[1] https://www.ruby-lang.org/en/

The original version of this chapter was revised. A correction to this chapter can be found at https://doi.org/10.1007/978-3-319-77809-9_10

**Electronic Supplementary Material:** The online version of this chapter (https://doi.org/10.1007/978-3-319-77809-9_6) contains supplementary material, which is available to authorized users.

© Springer International Publishing AG, part of Springer Nature 2018                    173
L. Brackney et al., *Building Energy Modeling with OpenStudio*,
https://doi.org/10.1007/978-3-319-77809-9_6

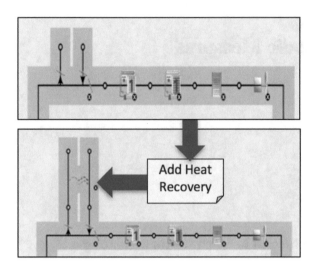

**Fig. 6.1** Measure used to add heat recovery to an HVAC system (https://bcl.nrel.gov/node/39440)

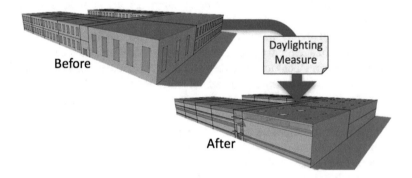

**Fig. 6.2** Measure used to apply a complete daylighting solution to a school (https://bcl.nrel.gov/node/39783)

In this example, a complete daylighting solution is applied to an arbitrary primary school Model with a single operation. This particular Measure is based on design guidance in the ASHRAE Advanced Energy Design Guide (AEDG) for K-12 schools[2] that includes:

- Removal of all existing fenestration,
- Addition of daylight and view glass to South-facing facades,
- Addition of daylight redirection devices (light shelves),
- Placement of daylight sensors and controls,
- Addition of skylights to spaces that benefit from top-lighting, and
- Addition of shading devices for glare control.

Note that the daylighting Measure "intelligently" traverses spaces within the Model, applying design elements surgically – e.g. skylights are added to a cafeteria

[2] ASHRAE (2011).

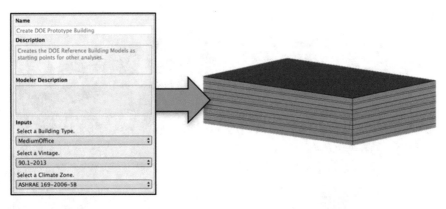

**Fig. 6.3**  DOE prototype building Measure (https://bcl.nrel.gov/node/83591) dialog and result

and gymnasium, but not in classrooms where top-lighting would interfere with projection Equipment. The Measure script embodies an expert modeler's knowledge of how to approach a good daylighting design; enabling a less experienced modeler to evaluate the Measure with far greater confidence than if they attempted to modify a Model by hand. Encapsulation of "best practice" for EE modeling is a powerful capability for increasing both speed and consistency of energy modeling.

Because Ruby is such a capable scripting language, OpenStudio Measures are not limited to modeling the application of EE technologies to buildings. They can also operate on an empty Model, using Ruby code along with the OpenStudio Standards Gem[3] to generate models from scratch. One example is the "Create DOE Prototype" Measure, which creates a Model procedurally from three Measure inputs: building type, vintage, and climate zone (Fig. 6.3). Behind scenes, this Measure (along with the Standards Gem) encapsulates a large amount of model input data and modeling heuristics corresponding to various building energy codes and standards. The net result from a user's perspective is that an incredibly sophisticated series of modeling operations are packaged within a deceptively simple interface.

Another type of OpenStudio Measure focuses on querying the data associated with the energy Model to produce customized reports or data exports. Although we didn't mention it, the standard OpenStudio reports we have reviewed in previous chapters are actually generated by a Measure, which enables easy customization and expansion. Other popular "reporting Measure" applications include interactive 3D visualizations for building geometry, automated Model quality checking, and HVAC psychrometric charts. Figure 6.4 presents html renderings for three reporting Measures.

Chapter 9 will cover the anatomy and creation of Measures in greater detail. The remainder of this chapter focuses on utilizing Measures to automate common modeling tasks.

---

[3] https://github.com/NREL/openstudio-standards. A Ruby Gem is a packaged library of Ruby code. The OpenStudio Standards Gem contains a collection of Ruby scripts that are useful for applying energy standards and input assumptions to models.

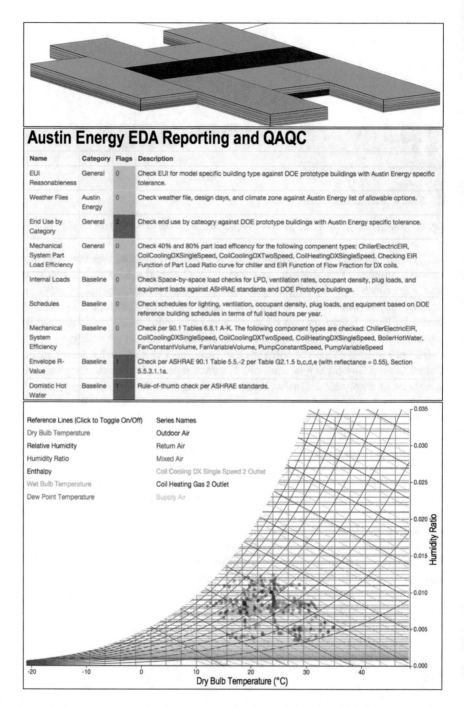

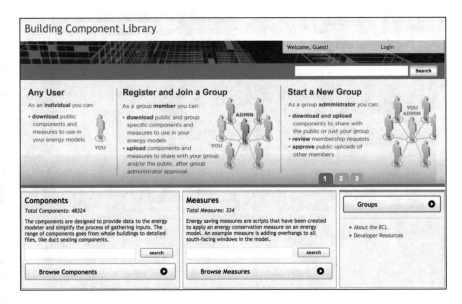

**Fig. 6.5**  Landing page of http://bcl.nrel.gov

## 6.2   Accessing and Using Measures

Since Measures are such an important part of the OpenStudio ecosystem, it stands to reason that ensuring they are easily accessible is an important feature of the platform. Measures are stored in an online repository called the Building Component Library (BCL).[4,5] The BCL can be accessed through a web browser, and includes integral search functionality to quickly locate content of interest. Figure 6.5 shows the landing page of http://bcl.nrel.gov.

An integrated search engine allows users to quickly identify content of interest – e.g. HVAC Measures (Fig. 6.6). The left side of the BCL window enables narrowing of search results using a combination of attributes.

Clicking on a specific entry brings up more information that can help the modeler determine if a particular Measure is right for the task. Figure 6.7 is the BCL entry for the ERV Measure we highlighted above. Eagle-eyed readers will note that this Measure has a specific URL (Uniform Resource Locator) that is unique to the Measure. Footnotes associated with Figs. 6.1, 6.2, 6.3, and 6.4 call out the respective URLs for the illustrated Measures, facilitating citation and sharing of specific content used in any given analysis. We'll revisit these Unique Identifiers (UIDs) along with the rest of a Measure's descriptive information in Chap. 9, but for the time being note the descriptive text and a ⬛ Button that allows the Measure to be downloaded.

---

[4] http://bcl.nrel.gov

[5] Fleming et al. (2012).

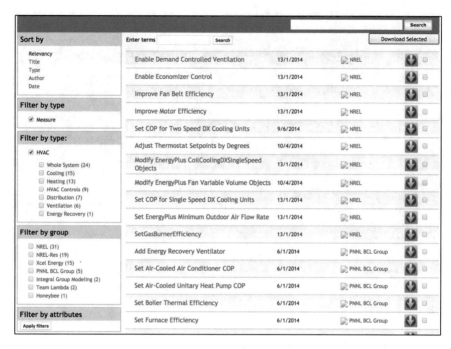

**Fig. 6.6** HVAC Measures on the BCL

**Fig. 6.7** BCL web page for ERV Measure

**Fig. 6.8**  Accessing the BCL from within the OpenStudio Application

## 6.3   Using Measures Within the OpenStudio Application

While the BCL web page provides a convenient means of identifying and downloading Measures, OpenStudio utilizes the BCL's API to integrate that search and download functionality directly in its applications. Not only does this improve convenience for the modeler, but it also allows OpenStudio applications to identify when a more recent version of the Measure is available for use. In the OpenStudio Application, the BCL is accessed from the "Components & Measures" menu (Fig. 6.8).

Once selected, a new BCL dialog pops up to allow search and selection of Measures. Measures already downloaded appear with a gray checkbox next to them. Clicking the checkbox next to one or more Measures and scrolling to the bottom of the list reveals a download Button that may be used to add additional content to a user's computer. In Fig. 6.9 we can see that the ERV Measure has already been downloaded and installed for use in an energy Model.

### 6.3.1   Applying a Measure Immediately to a Model

The OpenStudio Application offers two ways to use Measures: the "Apply Measure Now" menu option and via the Measure (▣) Tab. "Apply Measure Now" is located in the same menu used to open the BCL dialog (Fig. 6.8) and is used to select and apply a Measure to the active Model. As an example, consider the energy Model with a single Zone served by a packaged rooftop unit shown in Fig. 6.10.

Selecting the Apply Measure Now option, the user is first prompted to save their Model before proceeding to the Measure selection dialog (Fig. 6.11). Navigating the same categories used on the BCL, we can quickly locate the "AddEnergyRecoveryVentilator" Measure discussed previously. The OpenStudio icon and blue BCL text indicate this Measure has been downloaded from the BCL. Selecting that Measure updates the window with a description of the Measure and several Measure-specific inputs that can be customized. In this example, we have selected only the Packaged Rooftop Air Conditioner air loop for the Measure to act on and accept the default values for the remaining inputs related to heat exchanger efficiency.

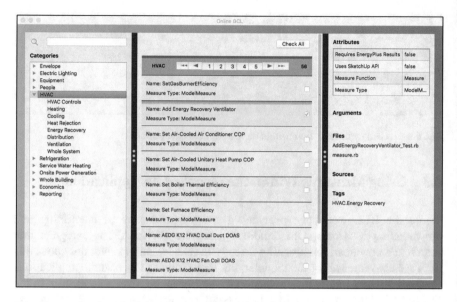

**Fig. 6.9** The BCL interface within the OpenStudio Application

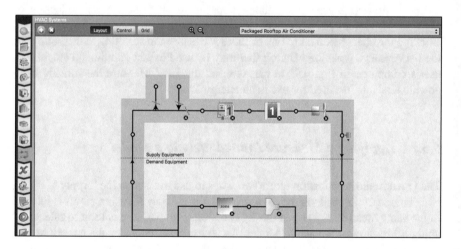

**Fig. 6.10** Model HVAC air loop prior to applying Measure

Clicking the "Apply Measure" Button transitions to a progress indicator while the Measure is applied to the Model. After a brief wait, a new dialog appears containing the Measure's "output." Just as Measure inputs vary, so do outputs. In this case, the ERV Measure reports what it did to the Model: reporting how many air loops it may have structurally modified along with Model parameters that have been set or altered (Fig. 6.12). This dialog would also report any errors that may have occurred when applying the Measure. This feedback allows the user to determine if the Measure behaved as expected, accepting those changes to the Model, or rejecting the modification with the Cancel Button.

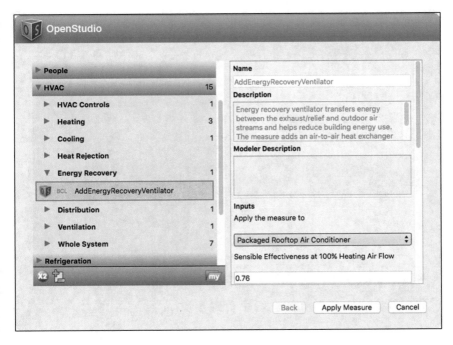

**Fig. 6.11**  Apply Measure Now window with ERV Measure selected

Figure 6.13 shows the OpenStudio HVAC (🔲) Tab after the Measure was applied, allowing us to visually confirm that the Measure added an ERV to the Packaged Rooftop Air Conditioner loop and that the parameters for the ERV were set as expected. Saving the Model once more will make this alteration permanent.

### 6.3.2   Adding Measures to an OpenStudio Application Workflow

The second method of applying Measures to a Model utilizes the Application's Measures (🔲) Tab (Fig. 6.14). The fundamental difference between this method and using "Apply Measure Now" is that the Measures (🔲) Tab does not permanently modify the building Model. Measures are applied to the Model and affect simulation results, but the Model itself is left untouched. This can be useful in seeing the effect of a Measure without fully committing to that as a design feature in the building Model.

Note that the Measures (🔲) Tab has sections for three "types" of Measures. They are OpenStudio Measures, EnergyPlus Measures, and Reporting Measures. Each has a unique icon, and on the right side of the Tab, we note that several of the available Measures have matching icons. To understand the meaning of these three categories, it is important to understand what OpenStudio does when we push the 🔲 Button on the Run Simulations (🔲) Tab. This run process or "workflow" is illustrated in Fig. 6.15.

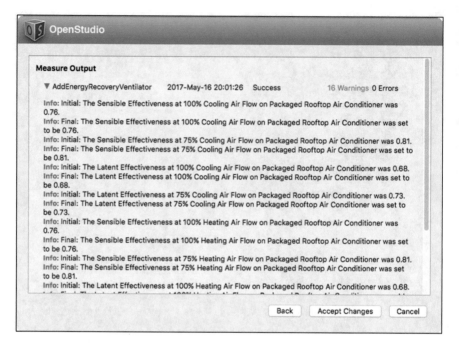

**Fig. 6.12** Report from ERV Measure after application

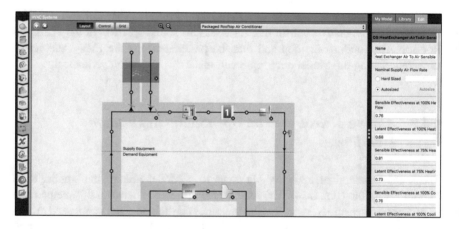

**Fig. 6.13** Model HVAC air loop after applying Measure

As one would expect, the process begins with the OSM. The first Measures to be applied to the Model are OpenStudio Measures, Measures that make full use of OpenStudio's Object Model and data inheritance. An OpenStudio Measure that removes an HVAC air loop removes not only the loop, but all of the objects within it and any associated Zone or Plant Loop Connections. Similarly, an OpenStudio Measure could alter a space type definition, simultaneously changing all of the space's underlying definitions for loads, occupancy, schedules, etc. OpenStudio Measures are the preferred means of modifying an OpenStudio Model.

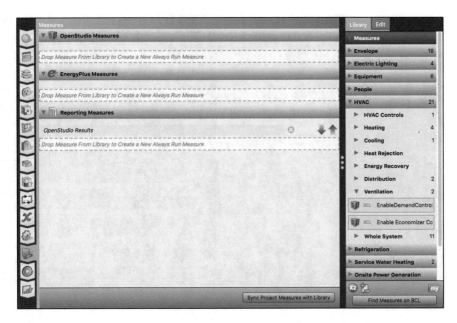

**Fig. 6.14** OpenStudio Application Measures Tab

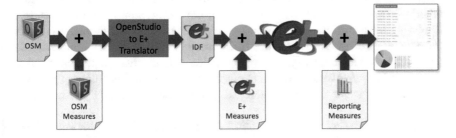

**Fig. 6.15** High level OpenStudio workflow with Measures

After all, OpenStudio Measures are applied to the Model, OpenStudio translates the Model from an OSM to an EnergyPlus IDF in preparation for actually running the simulation. At this point, EnergyPlus Measures can be applied to the IDF. EnergyPlus Measures are most often used to make specific alterations to the EnergyPlus IDF that may not yet be supported within the OpenStudio Object Model. These might include new, experimental, or esoteric EnergyPlus features that OpenStudio has yet to support. Examples at the time of writing include Measures that add utility tariffs to the Model or modify EnergyPlus Energy Management System (EMS) control logic. EnergyPlus Measures do not enjoy the benefits of the OpenStudio Object Model, and are best suited for simple, well-defined text additions or substitutions in the IDF file. Because of their limitations, EnergyPlus Measures should be used sparingly.

Following the application of any EnergyPlus Measures, OpenStudio invokes the EnergyPlus simulation engine itself. This produces both high level summaries

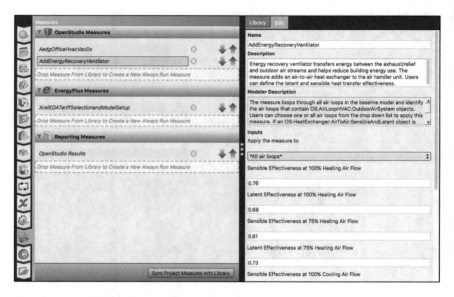

**Fig. 6.16** Measures added to a simulation workflow

of the simulation results in an EnergyPlus eplustbl.htm file, along with a SQLite database containing detailed time series results from the simulation. These (and other) EnergyPlus output files are the subjects of the final class of Measures, Reporting Measures. Reporting Measures can parse and format simulation results for a number of purposes ranging from creating simple Model summaries to applying automated quality check heuristics against the simulation results. Reporting Measures allow the user to customize the simulation output to produce static or interactive content suited to specific analysis needs or the target audience.

Figure 6.16 shows the Measures (■) Tab after several different Measures have been "dragged and dropped" into the three categories. In this example, we apply two OpenStudio Measures that modify the HVAC systems in the building, one EnergyPlus Measure that adds a utility rate tariff, and a Reporting Measure that produces the standard OpenStudio HTML report. As with the Apply Measure Now method, the user can select a Measure and modify all of its inputs on the right side of the window. It is also important to note the ▼▲ Buttons just to the right of the ⊠ Button that allows a Measure to be removed from the workflow. These arrows allow Measures to be reordered. This is important because Measures are generally non-commutative or order dependent. Consider this example where the "AEDGOfficeHvacVavDx" Measure operates on the Model before "AddEnergyRecoveryVentilator." In this case, the first Measure removes all existing HVAC systems from the Model, replacing them with VAV systems before ERVs are added. If the order is reversed, an ERV is first added to the Model's existing HVAC systems. This modification is then eradicated when those systems are replaced by VAV systems – quite likely not the intended result.

## 6.4   Introduction to Parametric Analysis

In practice, we wish to compare the performance of various building designs including individual EE measures or combinations thereof. The previous section showed how Measures might be applied to an OpenStudio Model to create Design Alternatives. While Measures make the process of altering the Model fast and repeatable compared to manually altering it, comparison of results would still be cumbersome; perhaps involving setup and management of multiple simulations and transcription of relevant outputs into a spreadsheet. Comparative analysis of Measure-based Design Alternatives is the role of another OpenStudio example application, the Parametric Analysis Tool (PAT).

PAT allows the user to quickly try out and compare specified combinations of measures, optimize designs, calibrate models, perform parametric sensitivity analysis, and much more. This section focuses on utilizing Measures to manually construct Design Alternatives that can be simulated using local computing resources. Chapter 7 delves more deeply into PAT's ability to explore design spaces utilizing a variety of sampling and search algorithms that can be run on dedicated computing hardware or in the cloud.

### 6.4.1   Starting PAT

When first launching PAT from the OpenStudio installer package the user has the option to create a new project or open an existing project (Fig. 6.17). An OpenStudio project represents a particular analysis comparing results from multiple models and resulting simulations.

To create a new project:

- Click the "Make New Project" Button.
- Type the name[6] for the project and click "Continue" as shown in Fig. 6.18.
- Browse to the directory where the project should be saved and click the "Open" Button (Fig. 6.19). PAT will create a new directory for the project at this location.

The user may also open an existing PAT project when first launching PAT, or from PAT's file menu. In either case a dialog will open to locate the Project directory. There is no file to select, just browse to the top-level of the project directory. Figure 6.20 illustrates selecting a previous project called "Office_HVAC."

As with most modern software applications, PAT's File menu also allows the user to create a new project, open an existing project, save a project, or save a copy of a project under a new name. Saving a copy of a project creates a new directory structure containing all of the project's resources.

---

[6] Important note – project names should <u>not</u> include spaces.

**Fig. 6.17**  Starting OpenStudio's Parametric Analysis Tool (PAT)

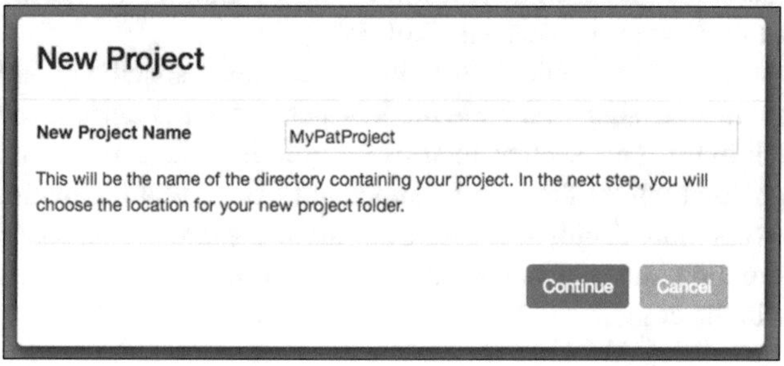

**New Project**

**New Project Name**      MyPatProject

This will be the name of the directory containing your project. In the next step, you will
choose the location for your new project folder.

Continue   Cancel

**Fig. 6.18**  PAT new project dialog

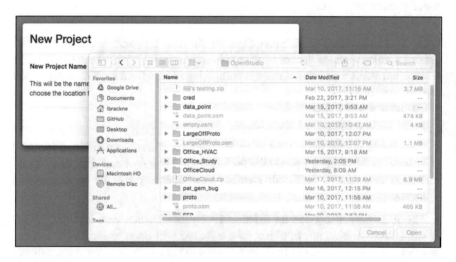

**Fig. 6.19**  Specifying a new PAT project's location

**Fig. 6.20** Opening the "Office_HVAC" project in PAT

### 6.4.2   Managing Measures in PAT

Like the OpenStudio Application, PAT's functionality is broken down into Tabs located along the left side of the window. The Tabs include:

- ⊡ **Analysis** – Used to specify the analysis mode, seed models, weather files, Measures, and Measure options;
- ◻ **Design alternatives** – Used to establish Design Alternatives;
- ◻ **Outputs** – Specifies outputs for algorithm-based analysis;
- ◻ **Run simulations** – Runs an analysis and manages the analysis server;
- ◻ **Compare results** – Used to view comparison reports; and
- ◻ **Analysis server** – Allows the user to view and interact with the underlying analysis server.

In general, a PAT workflow moves through the six vertical Tabs from top to bottom. PAT allows the user to specify Design Alternatives manually or automate the process using a selection of algorithms. The analysis mode is selected at the top of the Analysis (◻) Tab (Fig. 6.21). The choice of mode dictates whether the Design Alternatives (◻) or Outputs (◻) Tabs are used along with the types of computing resources (local or cloud) that may be selected on the Run Simulations (◻) Tab. This section focuses on PAT's "manual" mode. Algorithm-driven analyses are the subject of Chap. 7.

Just below the analysis mode selector are fields to identify a default "seed" Model and weather file. At this point, the weather file is self-explanatory. The seed Model is the initial Model that Measures in the project will be applied to. Clicking on the ☞ symbol allows the user to select the seed Model and weather file. Note that PAT supports specifying multiple seed and weather files that may all be used within a project for certain types of analyses.

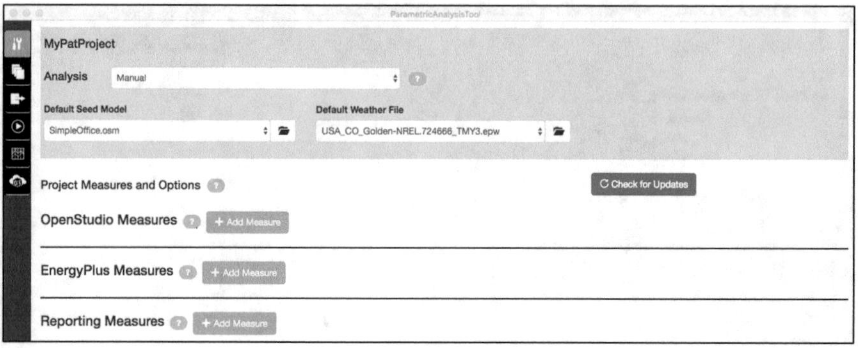

**Fig. 6.21** Specifying an analysis in PAT

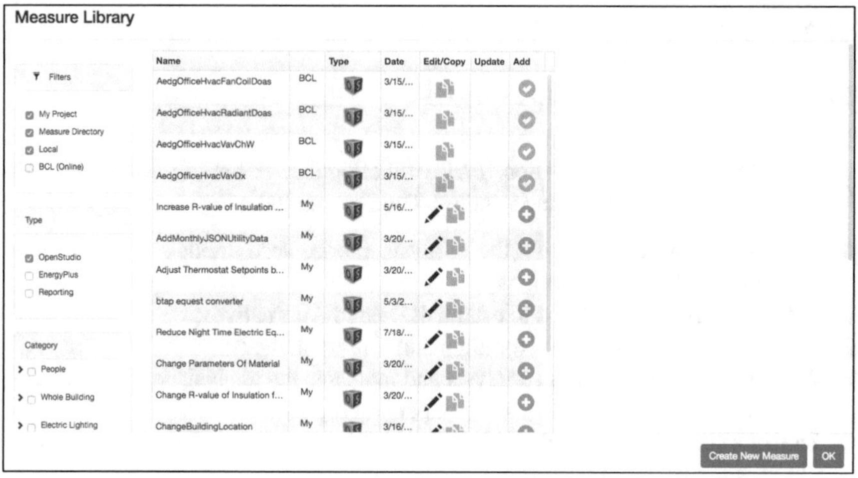

**Fig. 6.22** PAT's Measure Library dialog for the "Office_HVAC" project

As with the OpenStudio Application, PAT allows the selection of all three Measure types. Click the [+ Add Measure] Buttons by the OpenStudio, EnergyPlus, or Reporting Measure text to add Measures of that type to your project. The resulting Measure Library dialog is structurally similar to the BCL dialog we used in the OpenStudio Application. It provides the same search, download, and add functionality discussed in the previous section with a few additions. Figure 6.22 shows the dialog for an existing project that serves as a useful example in expanding our understanding of Measures.

The Measure Library dialog allows the user to filter by Measure location, type, category, and sub-category. When it first opens the dialog will have all locations checked except for BCL (Online). It will also have only one Measure type checked, based on which [+ Add Measure] Button was clicked. The category and sub-category options are identical to the categories discussed in the previous section and are helpful in locating a Measure more rapidly.

OpenStudio Measures are located in one of four distinct locations:

1. My Project – Measures that have already been added to the current PAT project. Such Measures will be denoted with a ⊘ under the "Add" column to indicate they've been added to the project.
2. Measure Directory – Measures stored in a user configurable "MyMeasures" directory on your computer. This topic will be discussed further in Chap. 9, but this directory generally will contain Measures that the user has customized or intends to customize. These Measures will have a "My" label next to the Measure name and may be added to the project by clicking the ⊘ Button.
3. Local – Includes any BCL Measures that may have already been downloaded. These Measures will have a "BCL" label next to the Measure name and may be added to the project by clicking the ⊘ Button.
4. BCL (Online) – Contains all publicly available Measures located in the online BCL that have yet to be downloaded.

Only Measures stored in Local or MyMeasures may be added to a PAT project, and only Measures added to the project may be used to create Design Alternatives for comparison. BCL Measures cannot be added until they are first downloaded – hence the reason the BCL selector is initially unchecked in the Measures Library dialog. Figure 6.23 shows the same project Measure Library dialog with the BCL selector turned on and the Envelope/Fenestration subcategory selected. Downloading a Measure from the BCL to Local is as easy as clicking the relevant ⊕ Button.

Once downloaded, PAT will check to see if an updated version of a Measure is available in the BCL each time the project is opened. Available updates will be indicated with a ▣ symbol in the "updates" column. Simply click the ▣ to download the latest version to your local library.

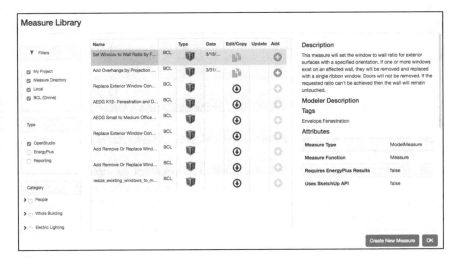

**Fig. 6.23** PAT's Measure Library with BCL fenestration content shown

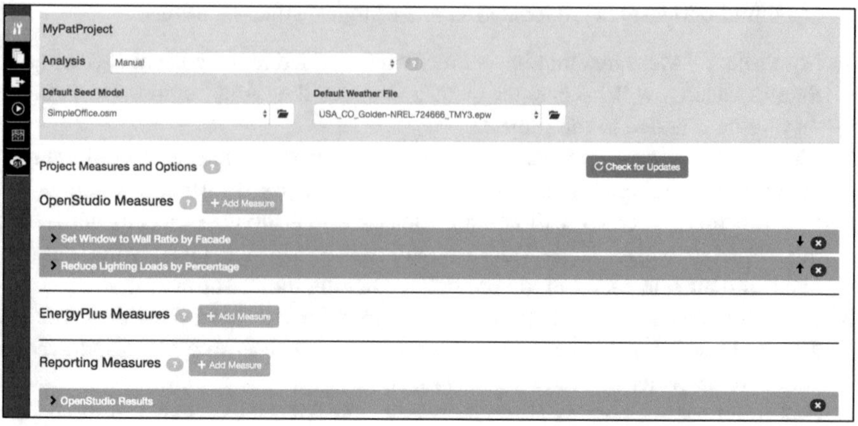

**Fig. 6.24** Three Measures added to a PAT project

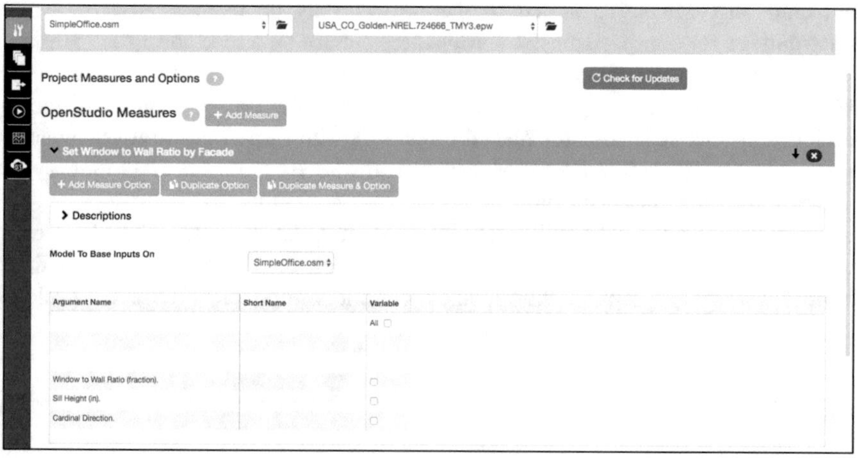

**Fig. 6.25** Set window to wall ratio by Façade Measure expanded

After making the desired selections with ⊙ and closing the BCL dialog, our Measures are now loaded into the project. In the example shown in Fig. 6.24, three measures have been added.

Recall from the previous section that the ordering of Measures matters. As in the OpenStudio Application, Measures in PAT run from top to bottom in the user interface, and they may be reordered using the ⬆ or ⬇ arrows at the right. The ⊗, may be used to delete a Measure from the project. Note that each Measure also has an ▶ just to its left. This is used to expand and collapse the Measure, allowing the user to specify Measure input arguments and more (Fig. 6.25).

Every Measure in a manual analysis project needs at least one Measure "option." An option describes a Measure and particular set of Measure arguments that will be applied to create a specific Design Alternative. For example, a generic fan efficiency

**Fig. 6.26**  A newly created Measure option

Measure would likely have an efficiency argument. Applying that generic Measure with a specific value for efficiency would constitute an option representing a specific fan product.

Clicking the ![Add Measure Option] Button adds a column to the right side of the grid for the Measure. Newly created options have a generic option name, description, and inherit any default values specified by the Measure. Figure 6.26 depicts a newly created "Set Window to Wall Ratio by Façade" Measure option with the Measure's default values.

The variable column and associated checkboxes are used to specify which arguments will vary across design options. Each option must be given a unique (and meaningful) name that will be referenced when constructing Design Alternatives. The option description field is free form and can be used to capture notes regarding the option that may be used by reporting Measures. In Fig. 6.27, three design options have been created with variable window to wall ratios.

The standard OpenStudio Results Measure must be added to every project as an option for each Design Alternative (Fig. 6.28). Manual project reports and most algorithmic workflows rely on outputs defined by this Measure to work properly. PAT will run without this Measure and an associated Measure option, but comparison results will not appear correctly.

### *6.4.3    Creating Design Alternatives in PAT*

The Design Alternatives (▢) Tab is used to create Design Alternatives consisting of a seed Model, weather file, and some combination of Measure options. Design alternatives may be based on multiple seed models, allowing for the inclusion of modeled alternatives for which no Measure exists. Buttons near the top of the window are used to create and copy individual alternatives. A Button to automatically create

**Fig. 6.27** Window to wall ratio Measure with three design options

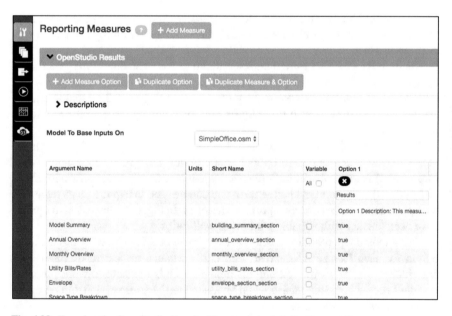

**Fig. 6.28** Ensuring the OpenStudio Results Measure is included with an option

a design with each Measure option applied independently is also available. Repeatedly pressing + Add Alternative creates sequentially named alternatives based on the default seed Model and weather file with no options defined.

Figure 6.29 illustrates this Tab with eight Design Alternatives specified. The Tab's grid layout allows for rapid inspection, alteration, and naming of alternatives. Clicking most of the fields brings up a menu that is automatically populated with available choices. Design alternative name and description fields are free from text. It is considered a best practice to enter meaningful descriptions in these fields to document the project.

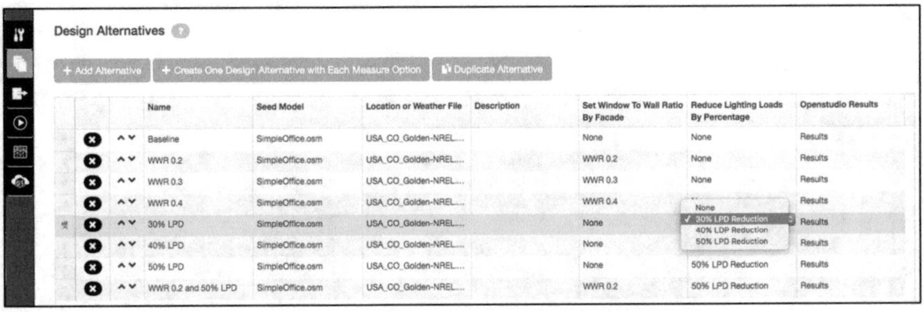

**Fig. 6.29** PAT's design alternatives Tab

Design alternatives may be deleted using the ⊗ Button on the left side of the screen. Rows may be rearranged manually using the ⌃⌄ Buttons or sorted alphabetically by clicking the grid headings. The order of Design Alternatives on this Tab has no bearing on simulation order or results reporting and is simply used as an aid to the modeler in defining the analysis.

> **Useful Tip**: Quickly set multiple options for a Measure by highlighting the first option, typing the first letter of the desired option, tapping the down arrow, and repeating. This is useful for quickly selecting things like the OpenStudio Results Measure in this example by repeatedly pressing the "R" and down keys.

### 6.4.4   Running an Analysis in PAT

PAT configures a "mini server" on your computer to perform local analysis. This is essentially the same server that is used to run large-scale cloud analyses that will be discussed in Chap. 7. PAT's server architecture enables it to easily migrate projects between scalable computing systems. The local server begins to start up as soon as you launch PAT and is usually ready to perform analysis within a minute. This is indicated by a Server Status ✓ message at the top of PAT's Run Simulations (▣) Tab shown in Fig. 6.30.

For a manual analysis, leave "Run Locally" selected. Pressing the Run Entire Workflow Button starts running simulations using all available CPU cores less one for PAT itself. In Fig. 6.31 we see that my meager laptop has four cores and is able to run three simulations at a time. As Design Alternative simulations (data points) are completed; associated rows are updated with run times, status, an active report selector ▥▾, and ⊘ indicators showing that the final Model and results are available for review.

PAT's status will continue updating to apprise the user of progress. Changing Tabs during active simulation is prevented to avoid accidentally altering an analysis mid-run. As data points complete, the user can inspect results from the Run

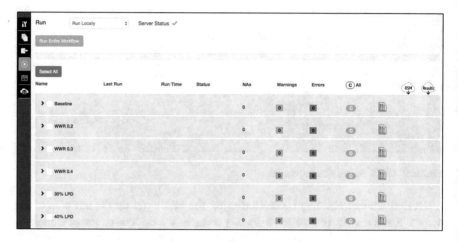

**Fig. 6.30** PAT's Run Tab indicates it is ready to simulate

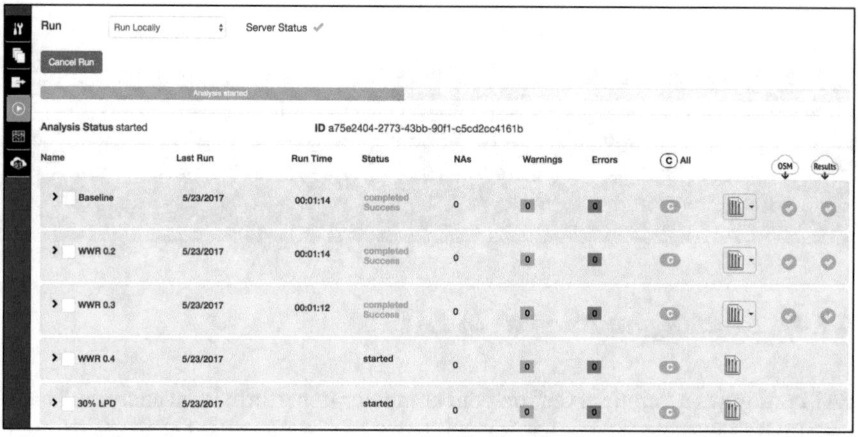

**Fig. 6.31** PAT analysis in progress

Simulations (▣) Tab by clicking on the report selector icon (Fig. 6.32). In this case, the user may select from the default EnergyPlus tabular report or the OpenStudio Results Measure that was added as an option for all the Design Alternatives.

Each data point may also be expanded by clicking on the ▹ Button on the left side of the Tab. Figure 6.33 shows the WWR 0.2 data point expanded revealing the Measure output messages for that alternative. Here we see that the window to wall ratio Measure changed the South-facing façades from 0.4 to 0.2 at no cost since we didn't enter a cost input. The lighting loads were unchanged since that option was set to "None" for this alternative. The OpenStudio Results Measure also states that it generated a report for the data point. **This output should be the first recourse for any user faced with a data point that has failed for one reason or another.**

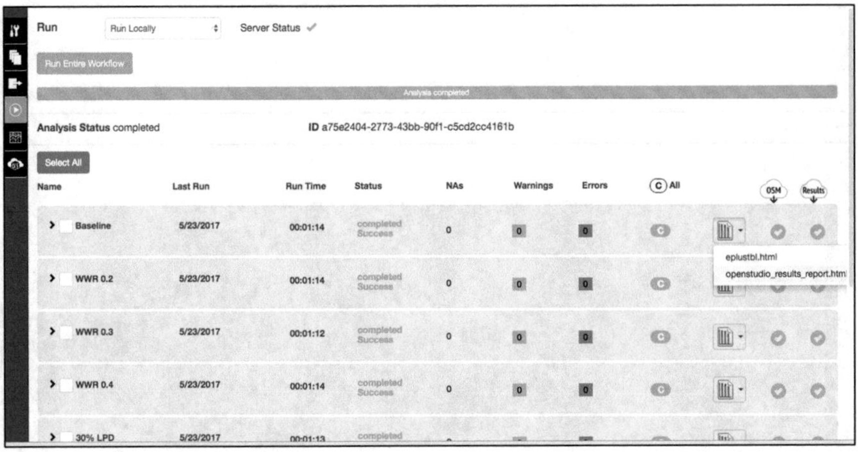

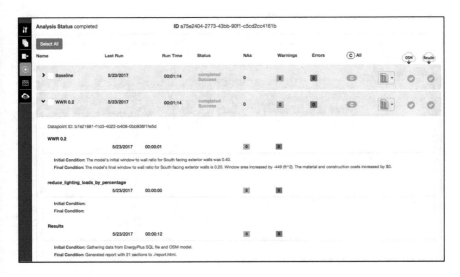

**Fig. 6.32** PAT completed analysis

**Fig. 6.33** An individual PAT data point expanded

In the event that a problem with a data point is corrected or the alternative is modified in some way, PAT allows individual data points to be re-run without repeating the entire set of simulations. A checkbox next to each Design Alternative's name allows the user to specify which data points are to be re-run. In Fig. 6.34 the "WWR 0.4" alternative has been selected for repeat simulation. Note that an additional Run Selected option appears at the top of the window when any checkboxes are ticked.

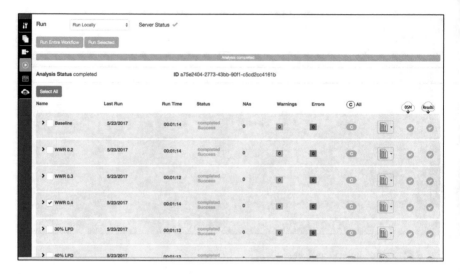

**Fig. 6.34** Re-running a single data point

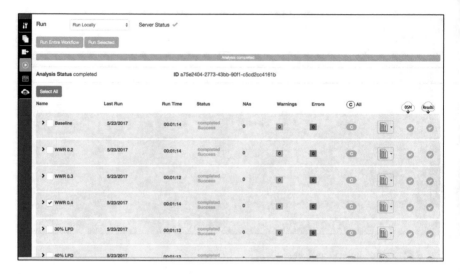

**Fig. 6.35** PAT summary table on the reports Tab

## 6.4.5   Comparing Results in PAT

PAT includes three built-in reports in its Compare Results (▦) Tab that help the user compare Design Alternatives. The default summary table compares consumption, demand, and economic metrics for all of the Design Alternatives. A selection field near the top of the table allows the user to specify which of the alternatives is to be used as the datum for performance comparison. Analysis results shown in subsequent rows are relative to the datum's modeled values. Figure 6.35 contains results for our example problem with the Baseline data point (seed Model + no Measures) used as a datum. Note that the cost columns are all

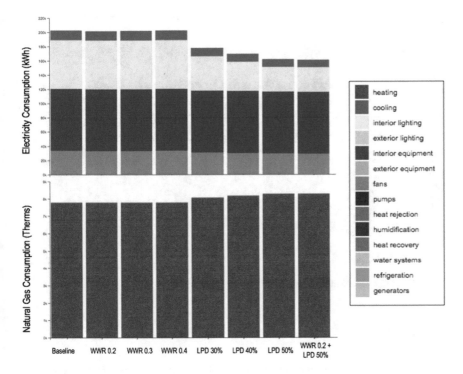

**Fig. 6.36** End user comparison chart

empty in this example because we did not specify a utility tariff Measure or any capital costs for the window to wall ratio or lighting power reduction changes to the building.

The reports selection field at the top of the window may also be used to produce end use stacked bar charts broken down by fuel type like the one shown in Fig. 6.36.

PAT includes a third reporting option that is used by professionals working in some utility programs that offer incentives for high performance buildings. That topic is beyond the scope of this textbook, but those interested can learn more about how PAT works with these programs at http://eda-pt.org.

### 6.4.6  The OpenStudio Analysis Server Tab

The Analysis Server (◻) Tab (Fig. 6.37) provides a view into PAT's underlying OpenStudio Server. While this Tab is generally not necessary for manual analyses, it provides our first glimpse at OpenStudio's deep capacity for large-scale analysis including sampling and optimization – the topics of Chap. 7.

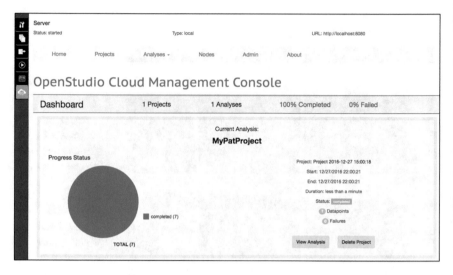

**Fig. 6.37** PAT's OpenStudio Cloud Management Console Tab

## 6.5   Checkpoint Nine: Introduction to Parametrics

For our next checkpoint exercise, let's revisit our Primary School Model from Chap. 4 as the subject of an introductory study using OpenStudio Measures. Use the initial PAT dialog window to create a new project called "MyPATSchoolProject." Remember that the directory you select indicates where the PAT project's directory will be created. In the Analysis (▣) Tab select Chap. 4 MyPrimarySchoolHVAC. osm as the default seed Model and the weather file you have used in previous exercises as shown in Fig. 6.38.

Unlike the OpenStudio Application, the OpenStudio Results Measure is not automatically included in a PAT workflow. Click the ⬚Add Measure Button in the Reporting Measures section to bring up the Measure Library dialog window (Fig. 6.39). Locate the Results Measure and click the ● Button to add it to your project.

Once the standard reporting Measure has been added to the project, create a Measure Option for it with the ⬚Add Measure Option Button. Name the new Option "Report." Also, make sure that at least one of the Result Measure's arguments is checked as a variable as shown in Fig. 6.40. This is not normally needed, but we are going to test run our seed Model by itself in PAT, which needs at least one variable to run an analysis. You can come back later and uncheck this as a variable (or not) once we have added proper EE Measures and variables to the project.

Proceed to the Design Alternatives (▢) Tab to create a baseline for our analysis. This data point will consist of our seed Model and the Results Measure. Click the ⬚Add Alternative to add the alternative shown in Fig. 6.41.

Save your project before moving on to PAT's Run Simulations (▣) Tab. Once you see Server Status ✓ at the top of the window, you can press the ⬚Run Entire Workflow Button to run the baseline data point. Figure 6.42 illustrates the baseline simulation in process.

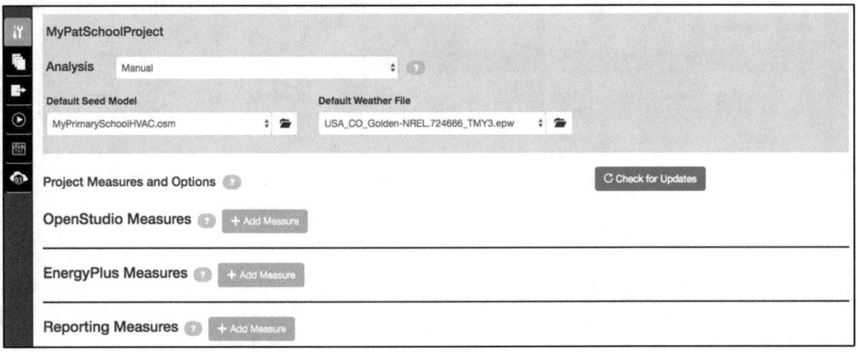

**Fig. 6.38**   Creating a new primary school PAT project

**Fig. 6.39**   Adding the OpenStudio Results Measure to the project

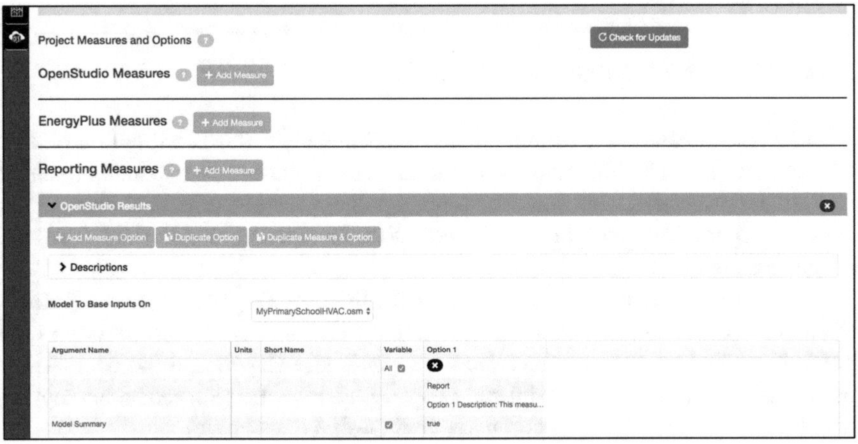

**Fig. 6.40**   Creating a Report Option for the project

**Fig. 6.41**  Creating the baseline Design Alternative in the project

**Fig. 6.42**  Running the project containing only the baseline

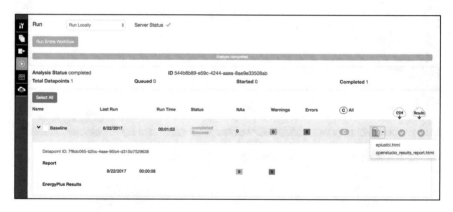

**Fig. 6.43**  Examining the baseline data point

When the simulation is complete, the Run Simulations ( ) Tab should look like the window shown in Fig. 6.43. Note that the data point can be expanded to show steps in the associated workflow. The reports pull down menu offers the EnergyPlus standard report along with the OpenStudio Results Measure output (Fig. 6.44) we added as an option for the data point.

The Compare Results ( ) Tab is shown as Fig. 6.45. This particular comparison table isn't particularly interesting since it only contains the single baseline data point. Nevertheless, we can see from the OpenStudio Report and the comparison table that the baseline Model did run properly, and we can proceed to adding EE Measures to perform a proper comparison of Design Alternatives.

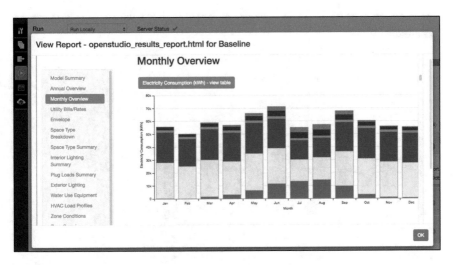

**Fig. 6.44** Reviewing the monthly end use breakdown for the baseline

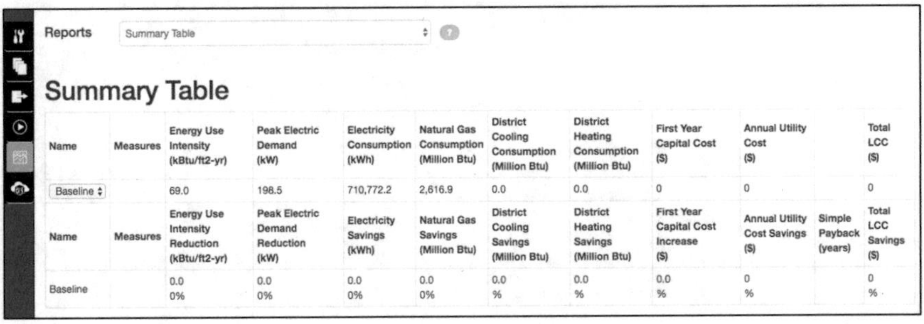

**Fig. 6.45** PAT comparison summary containing only the baseline

To add additional Measures and Design Alternatives, first return to the Analysis (■) Tab and click the [+ Add Alternative] Button next to the OpenStudio Measures section of the window. Let's first begin by downloading some HVAC system Measures from the BCL. Check the search boxes as shown in Fig. 6.46 to limit our search to whole system HVAC OpenStudio Measures available on the BCL.

Click the ⊕ Button next to the "AedgK12HvacDualDuctDoas" Measure to download it to our Local Measures Library. Using the current filter settings, that Measure appears to vanish. Never fear! Adding the Local Library to our filter reveals that it was downloaded and is available to add to the project using the ○ Button (Fig. 6.47).

Download the "AedgK12HvacFanCoilDoas" and "AedgK12HvacGshpDoas" Measures and add all three to your project (Fig. 6.48).

Dismiss the Measure Library browser and expand the AedgK12HvaDualDuctDoas Measure (Fig. 6.49). Add a Measure Option as shown in Fig. 6.50. Be sure to give it a meaningful name. Also note that this particular Measure has three arguments. We will set the total cost argument to be $150,000 for the purpose of this exercise. The other arguments should be left to their default values.

## Measure Library

| | Name | | Type | Date | Edit/Copy... | Update... | Add |
|---|---|---|---|---|---|---|---|
| ▼ Filters | AEDG K12 HVAC Dual Duct DOAS | BCL | | | ⊕ | | ⊕ |
| ☐ My Project | AEDG K12 HVAC Fan Coil DOAS | BCL | | | ⊕ | | ⊕ |
| ☐ Measure Directory | AEDG K12 HVAC GSHP DOAS | BCL | | | ⊕ | | ⊕ |
| ☐ Local | | | | | | | |
| ☑ BCL (Online) | AEDG Office HVAC ASHP with D... | BCL | | | ⊕ | | ⊕ |
| | AEDG Office HVAC Fan Coil DOAS | BCL | | | ⊕ | | ⊕ |
| **Type** | AEDG Office HVAC Radiant with ... | BCL | | | ⊕ | | ⊕ |
| ☑ OpenStudio | AEDG Office HVAC VAV with Chill... | BCL | | | ⊕ | | ⊕ |
| ☐ EnergyPlus | AEDG Office HVAC VAV with DX ... | BCL | | | ⊕ | | ⊕ |
| ☐ Reporting | AEDG Office HVAC WSHP with D... | BCL | | | ⊕ | | ⊕ |
| | Rooftop Unit | BCL | | | ⊕ | | ⊕ |
| **Category** | | | | | | | |
| ❯ ☐ People | GLHEPro GFunction Import | BCL | | | ⊕ | | ⊕ |
| ❯ ☐ Whole Building | Enable Ideal Air Loads For All Zon... | BCL | | | ⊕ | | ⊕ |
| ❯ ☐ Electric Lighting | GSHP with DOAS (More Design P... | BCL | | | ⊕ | | ⊕ |
| ❯ ☐ Envelope | VRFwithDOAS | BCL | | | ⊕ | | ⊕ |
| ❯ ☐ Equipment | WSHP with DOAS (More Design P... | BCL | | | ⊕ | | ⊕ |
| ❮ ☐ HVAC | add_aqua_therm_system | BCL | | | ⊕ | | ⊕ |
| ☐ Heating | Chilled Beam with DOAS | BCL | | | ⊕ | | ⊕ |
| ☐ Heat Rejection | | | | | | | |
| ☐ Ventilation | Replace HVAC with WSHP and D... | BCL | | | ⊕ | | ⊕ |
| ☐ Energy Recovery | | | | | | | |
| ☐ Distribution | Replace HVAC with GSHP and D... | BCL | | | ⊕ | | ⊕ |
| ☐ HVAC Controls | Add a PSZ-HP to each zone | BCL | | | ⊕ | | ⊕ |
| ☐ Cooling | | | | | | | |
| ☑ Whole System | | | | | | | |
| ❯ ☐ Refrigeration | | | | | | | |

**Fig. 6.46** Finding EE Measures for the project

Create Measure Options for the other two HVAC Measures as shown in Figs. 6.51 and 6.52. Be sure to provide appropriate Option names and costs for each system.

The last Measure we will add to our HVAC system comparison is a utility tariff or rate. Along with each system's capital cost, which we entered above, the utility rate will be used to calculate the relative cost of each Design Alternative in addition to energy savings. Utility rates are implemented as EnergyPlus Measures, so be sure to click the ⊞ Add Alternative Button next to the correct Measure category. Download and add the "XcelEDATariffSelectionandModelSetup" Measure to the Project (Fig. 6.53). This Measure adds energy rates charged by Xcel Energy, the utility that will serve our school in the Denver Colorado area.

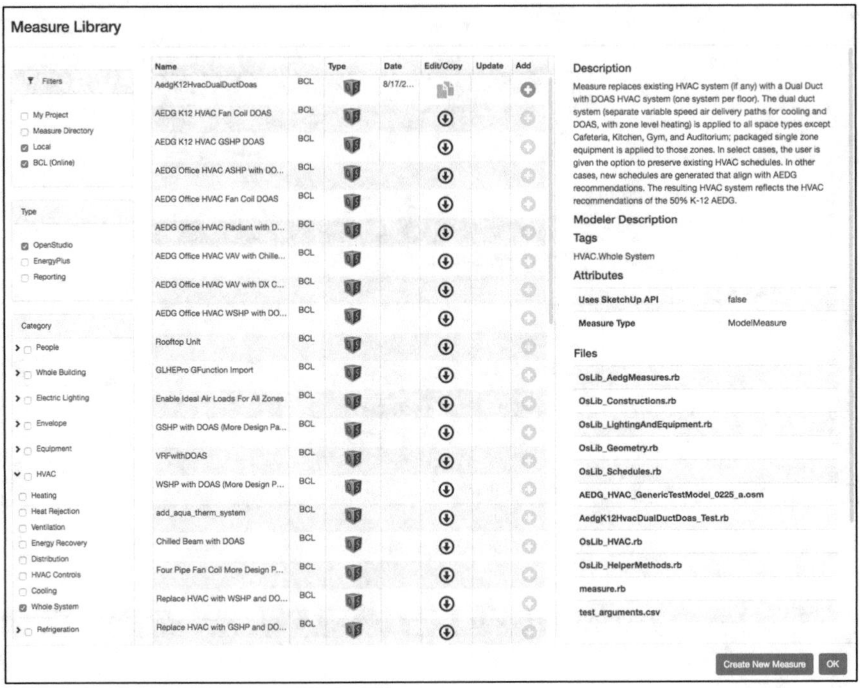

**Fig. 6.47**   Downloading the AEDGK12HVACDualDuctDoas Measure

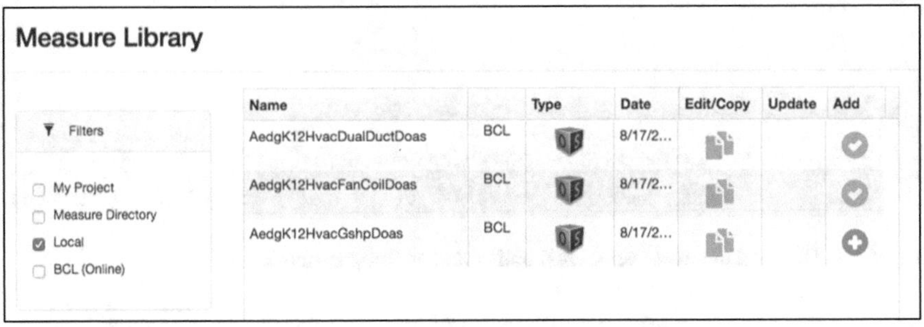

**Fig. 6.48**   Adding three AEDG K-12 HVAC Measures to the project

Finally, create a Measure Option for the utility rate as shown in Fig. 6.54. Note that this Measure contains two input arguments. Each is a list of choices for the electricity and gas tariffs appropriate for our building. Select "Commercial" and "Small CG" for the electricity and gas rates respectively.

Defining the additional Design Alternatives is performed on the Design Alternatives (⬛) Tab. Add three additional alternatives as shown in Fig. 6.55. It's worth taking the time to type in proper names for each alternative so that the comparison report will be more meaningful. Relying on PAT's automatic naming scheme

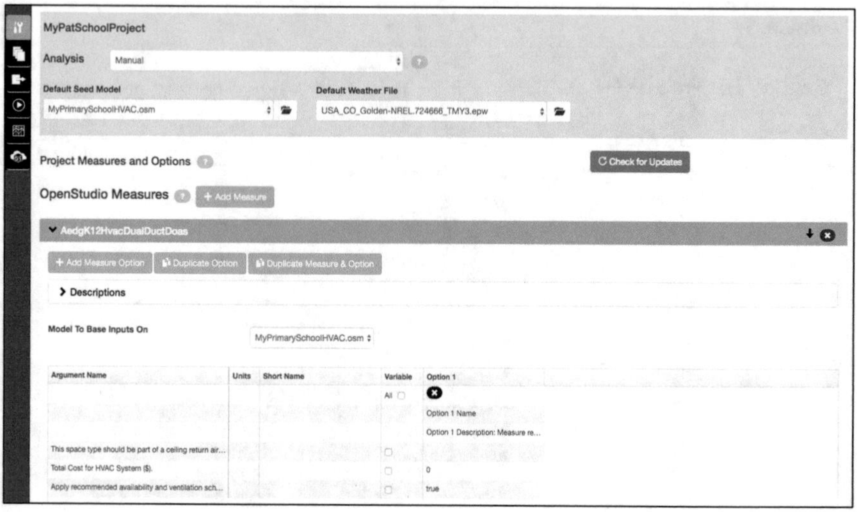

**Fig. 6.49** Defaults for a new dual duct DOAS option

**Fig. 6.50** Defining the dual duct DOAS option and entering system cost

of Alternative 1, Alternative 2, etc. may be fast, but it's not always so easy to remember what was included in Alternative 13 after you've run a bunch of simulations!

Also, don't forget to assign the Utility Tariff and Report Options to each Design Alternative. One way to quickly do this is to check the Baseline Alternative and then use the [Duplicate Alternative] Button to make copies. This preserves all of the Options from the copied data point, allowing you to simply change different Options in subsequent points.

Once the Alternatives have been defined, save your Project. It's time to kick off our simulations in the Run Simulations (▣) Tab. You can save a bit of time by checking just the new data points and clicking [Run Selected], or you can click [Run Entire Workflow] to run everything. Figure 6.56 shows the Run Simulations (▣) Tab with simulations in progress.

**Fig. 6.51**   Creating a fan coil DOAS Measure option

**Fig. 6.52**   Creating a GSHP DOAS Measure option

While simulations are running, it's worth taking a moment to look at the structure of a typical PAT Project directory (Fig. 6.57). As data points run and complete, subdirectories with rather cryptic names appear. Each of these directories corresponds to one of our data points, and includes the data point's OSM, results, etc. The data_point.zip also includes the detailed EnergyPlus files related to the data point.

Compare Fig. 6.57 with Fig. 6.58 and note that one of the data point IDs shown in PAT matches one of these directories. If you should ever need access to specific files associated with a given Design Alternative – be it a Model, results, or EnergyPlus files for troubleshooting; start by finding the data point ID, then going to the appropriate subdirectory in the PAT Project's *localResults* directory.

While we have that data point "open," take a moment to look at the diagnostic messages from each Measure in the data point's workflow. Note that both the dual duct DOAS and fan coil DOAS Measures were "skipped," but that the GSHP DOAS Measure removed the old HVAC system from the Model and replaced it with the new system. The utility tariff was also applied to the Model. As stated earlier in the chapter, these messages can be incredibly valuable in verifying that a workflow was executed as expected and troubleshooting problems when things don't go as planned.

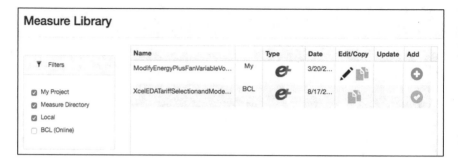

**Fig. 6.53** Adding the Xcel Energy utility rate tariff to the project

**Fig. 6.54** Creating a utility rate Measure option

**Fig. 6.55** Adding three additional Design Alternatives to the project

Now that we have multiple Design Alternatives, the Compare Results (▦) Tab is more meaningful. Selected the Baseline data point as the basis for comparison and you should see a performance summary similar to Fig. 6.59. Because we have added capital and utility costs to our data points, PAT is able to compare not only energy performance, but also the simple payback and total life cycle cost of the proposed HVAC alternatives.

Note that the Ground Source Heat Pump alternative contains data in both the district heating and cooling columns while the other alternatives do not. Does this make sense? What is the best system choice? Is there a clear winner?

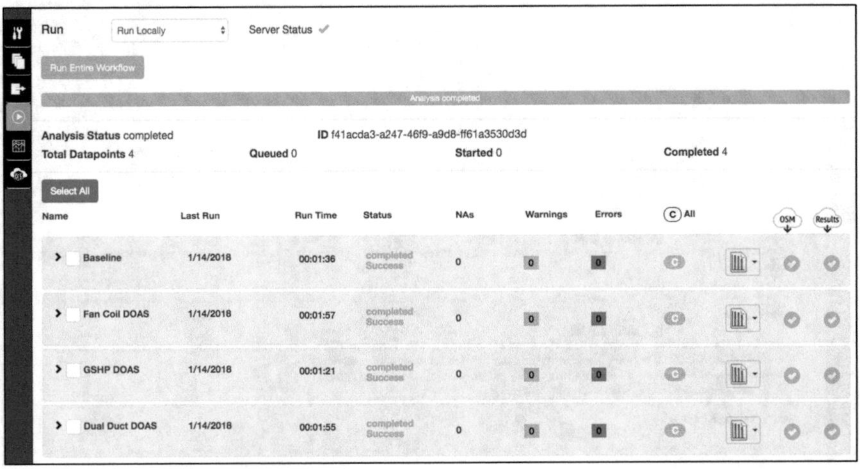

**Fig. 6.56**   Simulating the four Design Alternatives

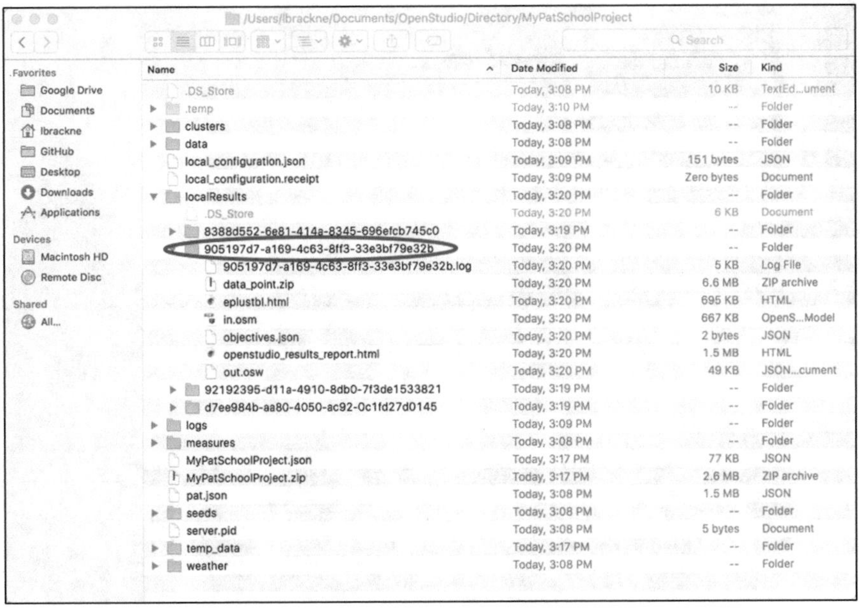

**Fig. 6.57**   PAT Project directory structure

Next, let's add a few more EE Measures into the mix. Return to the Analysis (□) Tab and add the Measures shown in Fig. 6.60. You will need to locate and download these from the BCL.

Use Table 6.1 for Option names and arguments as you set up your Measure Options.

**Fig. 6.58** Reviewing the Measure actions that produced the GSHP DOAS data point

## Summary Table

| Name | Measures | Energy Use Intensity (kBtu/ft2-yr) | Peak Electric Demand (kW) | Electricity Consumption (kWh) | Natural Gas Consumption (Million Btu) | District Cooling Consumption (Million Btu) | District Heating Consumption (Million Btu) | First Year Capital Cost ($) | Annual Utility Cost ($) | | Total LCC ($) |
|------|----------|-----------|-----------|-----------|-----------|-----------|-----------|-----------|-----------|--|-----------|
| Baseline ⇕ | | 69.0 | 198.9 | 710,599.7 | 2,613.8 | 0.0 | 0.0 | 0 | 89,965 | | 1,578,071 |

| Name | Measures | Energy Use Intensity Reduction (kBtu/ft2-yr) | Peak Electric Demand Reduction (kW) | Electricity Savings (kWh) | Natural Gas Savings (Million Btu) | District Cooling Savings (Million Btu) | District Heating Savings (Million Btu) | First Year Capital Cost Increase ($) | Annual Utility Cost Savings ($) | Simple Payback (years) | Total LCC Savings ($) |
|------|----------|-----------|-----------|-----------|-----------|-----------|-----------|-----------|-----------|-----------|-----------|
| GSHP DOAS | • GSHP DOAS | 24.9 36% | -10.7 -5% | 35,305.4 5% | 2,276.7 87% | -344.9 -∞% | -232.6 -∞% | 250,000.0 ∞% | 19,984 22% | 12.5 | 115,291 7% |
| Fan Coil DOAS | • Fan Coil DOAS | 29.7 43% | -7.7 -4% | 57,376.6 8% | 1,975.2 76% | 0.0 % | 0.0 % | 100,000.0 ∞% | 19,913 22% | 5.0 | 258,345 16% |
| Dual Duct DOAS | • Dual Duct DOAS | 12.4 18% | -24.0 -12% | -93,159.2 -13% | 1,225.5 47% | 0.0 % | 0.0 % | 150,000.0 ∞% | -443 0% | -338.4 | -147,956 -9% |

**Fig. 6.59** Comparing the three Design Alternatives against the baseline

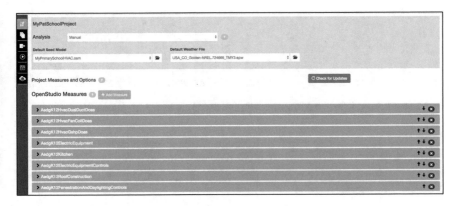

**Fig. 6.60**  Adding additional Measures to the project

**Table 6.1**  EE Measure Option names and arguments

| Measure name | Option name | Arguments |
|---|---|---|
| AedgK12ElectricEquipment | Improve electric equipment efficiency | $2 per equipment |
| AedgK12Kitchen | Improve kitchen efficiency | $40,000 |
| | | 200 students |
| AedgK12ElectricEquipmentControls | Improve electric equipment controls | $1500 |
| AedgK12RoofConstructions | Improved roof construction | $1 per ft$^2$ insulated |
| | | $0 per ft$^2$ solar |
| AedgK12FenestrationAndDaylightingControls | Fenestration and daylight controls | $1 per ft$^2$ daylight |
| | | $1 per ft$^2$ view |
| | | $2 per ft$^2$ skylight |
| | | $0.5 per ft$^2$ shading |
| | | $2 per ft$^2$ light shelf |

On the Design Alternatives (⬚) Tab, add a new Option that combines the GSHP system with all of the new EE Measures we just added. Again, don't forget to make sure that this new alternative includes the utility tariff and standard report as shown in Fig. 6.61.

Run the additional Design Alternative on the Run Simulations (⬚) Tab and use the Compare Results (⬚) Tab to compare the results with the previous Alternatives. Figures 6.62 and 6.63 compare the additional data point with the previous HVAC alternatives. Do these results make sense?

To conclude this exercise, recall that none of the Measures we have added to our school project included more than one Option. Let's get a bit of practice in with multiple Options by adding the "Reduce Lighting Loads by Percentage" Measure to our Project as shown in Fig. 6.64.

| | | Name | Seed Model | Location or Weather File | Description | Aedg K 12 Hvac Dual Duct Doas | Aedg K 12 Hvac Fan Coil Doas | Aedg K 12 Hvac Gshp Doas | Aedg K 12 Electric Equipment | Aedg K 12 Kitchen | Aedg K 12 Electric Equipment Controls | Aedg K 12 Roof Construction | Aedg K 12 Fenestration And Daylighting Controls | Xcel Eda Tariff Selectionand Model Setup | Openstudio Results |
|---|---|---|---|---|---|---|---|---|---|---|---|---|---|---|---|
| ⊗ | ^ ∨ | Baseline | MyPrimarySch... | USA_CO_Gold... | | None | None | None | None | None | None | None | None | Utility Tariff | Report |
| ⊗ | ^ ∨ | Dual Duct DOAS | MyPrimarySch... | USA_CO_Gold... | | Dual Duct DO... | None | None | None | None | None | None | None | Utility Tariff | Report |
| ⊗ | ^ ∨ | Fan Coil DOAS | MyPrimarySch... | USA_CO_Gold... | | None | Fan Coil DOAS | None | None | None | None | None | None | Utility Tariff | Report |
| ⊗ | ^ ∨ | GSHP DOAS | MyPrimarySch... | USA_CO_Gold... | | None | None | GSHP DOAS | None | None | None | None | None | Utility Tariff | Report |
| ⊗ | ^ ∨ | GSHP with Additional M... | MyPrimarySch... | USA_CO_Gold... | | None | None | GSHP DOAS | Improve Electr... | Improve Kitch... | Improve Electr... | Improved Roo... | Fenestration a... | Utility Tariff | Report |

**Fig. 6.61**  Defining a Design Alternative with GSHP and additional EE Measures

| Name | Measures | Energy Use Intensity (kBtu/ft2-yr) | Peak Electric Demand (kW) | Electricity Consumption (kWh) | Natural Gas Consumption (Million Btu) | District Cooling Consumption (Million Btu) | District Heating Consumption (Million Btu) | First Year Capital Cost ($) | Annual Utility Cost ($) | | Total LCC ($) |
|---|---|---|---|---|---|---|---|---|---|---|---|
| Baseline | | 69.0 | 198.9 | 710,599.7 | 2,613.8 | 0.0 | 0.0 | 0 | 89,965 | | 1,578,071 |

| Name | Measures | Energy Use Intensity Reduction (kBtu/ft2-yr) | Peak Electric Demand Reduction (kW) | Electricity Savings (kWh) | Natural Gas Savings (Million Btu) | District Cooling Savings (Million Btu) | District Heating Savings (Million Btu) | First Year Capital Cost Increase ($) | Annual Utility Cost Savings ($) | Simple Payback (years) | Total LCC Savings ($) |
|---|---|---|---|---|---|---|---|---|---|---|---|
| GSHP with Additional Measures | • GSHP DOAS<br>• Improve Electric Equipment Efficiency<br>• Improve Kitchen Efficiency<br>• Improve Electric Equipment Controls<br>• Improved Roof Construction<br>• Fenestration and Daylight Controls | 30.9 45% | -0.6 0% | 146,779.3 21% | 2,392.6 92% | -312.0 ~% | -320.8 ~% | 377,701.9 | 31,807 35% | 11.9 | 197,687 13% |
| GSHP DOAS | • GSHP DOAS | 24.9 36% | -10.7 -5% | 35,305.4 5% | 2,276.7 87% | -344.9 ~% | -232.6 ~% | 250,000.0 | 19,984 22% | 12.5 | 115,291 7% |
| Fan Coil DOAS | • Fan Coil DOAS | 29.7 43% | -7.7 -4% | 57,376.6 8% | 1,975.2 76% | 0.0 % | 0.0 % | 100,000.0 | 19,913 22% | 5.0 | 258,345 16% |
| Dual Duct DOAS | • Dual Duct DOAS | 12.4 18% | -24.0 -12% | -93,159.2 -13% | 1,225.5 47% | 0.0 % | 0.0 % | 150,000.0 | -443 0% | -338.4 | -147,956 -9% |

**Fig. 6.62**  Comparing the four Design Alternatives to the baseline

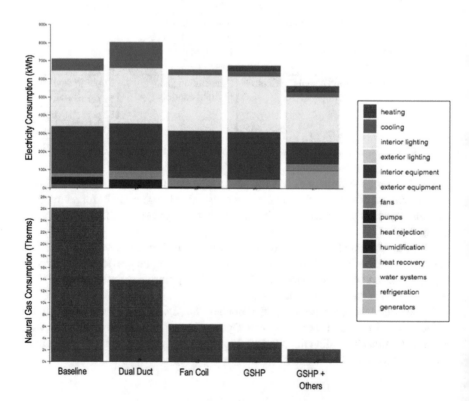

**Fig. 6.63**  Comparing end use breakdowns for the baseline and Design Alternatives

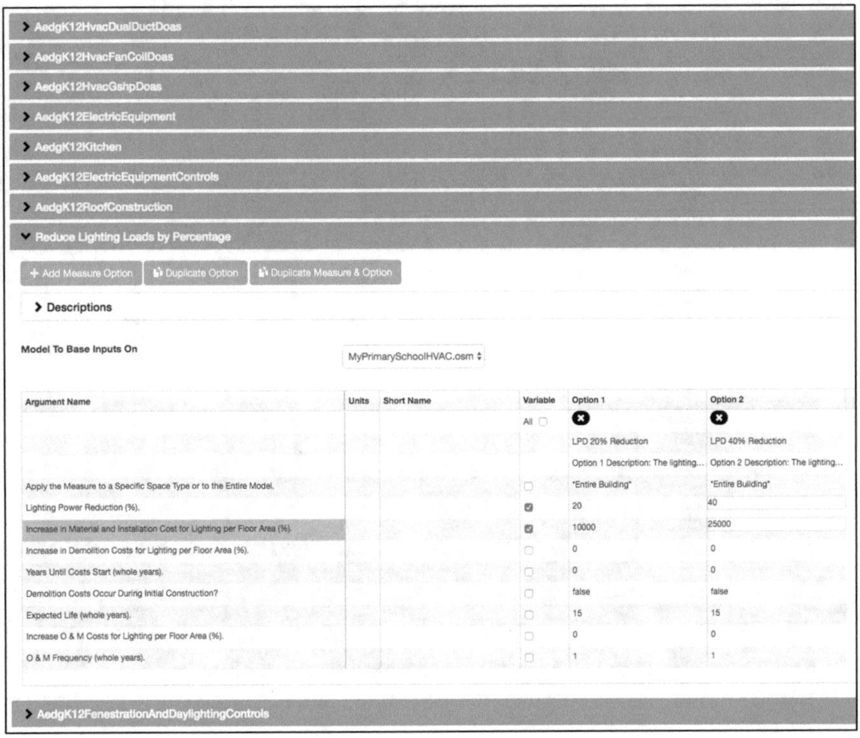

**Fig. 6.64** Adding the "Reduce Lighting Loads by Percentage" Measure with two options

Note that arguments we intend to vary across an option must be checked as variables or PAT will not allow us to change their values. It is also important that each Option be given a unique and descriptive name. Lastly, note that we have placed the lighting load reduction Measure ahead of the fenestration and daylighting controls Measure. This ensures that we modify all light fixtures in each Space before daylighting controls are applied.

Create two more Design Alternatives that include the GSHP System, all of the previous EE Measures, and either the 20% or 40% LPD reduction Option. Running those additional points should produce the comparison shown in Figs. 6.65 and 6.66.

Have our changes to LPD had the expected results in the last two Design Alternatives? What other Measures might we consider for our school, and in what combinations? Now that you have a reasonable understanding of how Measures work with PAT and access to additional Measures from the BCL, consider exploring the school's design for additional energy and cost savings.

| Name | Measures | Energy Use Intensity (kBtu/ft2-yr) | Peak Electric Demand (kW) | Electricity Consumption (kWh) | Natural Gas Consumption (Million Btu) | District Cooling Consumption (Million Btu) | District Heating Consumption (Million Btu) | First Year Capital Cost ($) | Annual Utility Cost ($) | | Total LCC ($) |
|---|---|---|---|---|---|---|---|---|---|---|---|
| Baseline | | 69.0 | 198.9 | 710,599.7 | 2,613.8 | 0.0 | 0.0 | 0 | 89,965 | | 1,578,071 |

| Name | Measures | Energy Use Intensity Reduction (kBtu/ft2-yr) | Peak Electric Demand Reduction (kW) | Electricity Savings (kWh) | Natural Gas Savings (Million Btu) | District Cooling Savings (Million Btu) | District Heating Savings (Million Btu) | First Year Capital Cost Increase ($) | Annual Utility Cost Savings ($) | Simple Payback (years) | Total LCC Savings ($) |
|---|---|---|---|---|---|---|---|---|---|---|---|
| LPD 40% | • GSHP DOAS<br>• Improve Electric Equipment Efficiency<br>• Improve Kitchen Efficiency<br>• Improve Electric Equipment Controls<br>• Improved Roof Construction<br>• LPD 40% Reduction<br>• Fenestration and Daylight Controls | 38.2 55% | 22.6 11% | 249,384.6 35% | 2,372.3 91% | -219.3 -∞% | -212.0 -∞% | 377,701.9 -∞% | 42,057 47% | 9.0 | 376,077 24% |
| LPD 20% | • GSHP DOAS<br>• Improve Electric Equipment Efficiency<br>• Improve Kitchen Efficiency<br>• Improve Electric Equipment Controls<br>• Improved Roof Construction<br>• LPD 20% Reduction<br>• Fenestration and Daylight Controls | 34.7 50% | 10.5 5% | 198,537.2 28% | 2,382.6 91% | -262.8 -∞% | -261.7 -∞% | 377,701.9 -∞% | 36,976 41% | 10.2 | 287,660 18% |
| GSHP with Additional Measures | • GSHP DOAS<br>• Improve Electric Equipment Efficiency<br>• Improve Kitchen Efficiency<br>• Improve Electric Equipment Controls<br>• Improved Roof Construction<br>• Fenestration and Daylight Controls | 30.9 45% | -0.6 0% | 146,779.3 21% | 2,392.6 92% | -312.0 -∞% | -320.8 -∞% | 377,701.9 -∞% | 31,807 35% | 11.9 | 197,687 13% |
| GSHP DOAS | • GSHP DOAS | 24.9 36% | -10.7 -5% | 35,305.4 5% | 2,276.7 87% | -344.9 -∞% | -232.6 -∞% | 250,000.0 -∞% | 19,984 22% | 12.5 | 115,291 7% |
| Fan Coil DOAS | • Fan Coil DOAS | 29.7 43% | -7.7 -4% | 57,376.6 8% | 1,975.2 76% | 0.0 % | 0.0 % | 100,000.0 -∞% | 19,913 22% | 5.0 | 258,345 16% |
| Dual Duct DOAS | • Dual Duct DOAS | 12.4 18% | -24.0 -12% | -93,159.2 -13% | 1,225.5 47% | 0.0 % | 0.0 % | 150,000.0 -∞% | -443 0% | -338.4 | -147,956 -9% |

**Fig. 6.65**  Comparing the two additional Design Alternatives with the baseline

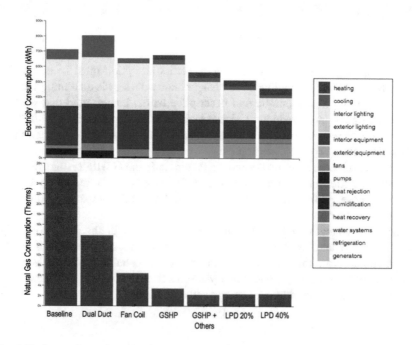

**Fig. 6.66**  Comparing end user breakdowns for baseline and alternatives

## 6.6    Additional Exercises

1) Recommended additional exercises involving the Checkpoint Nine Model include application of additional Measures within the PAT. The BCL contains a large number of Measures related to envelope, lighting, loads, systems and more. Spend some time navigating the available Measures and create additional Design Alternatives for further comparison with the Checkpoint Nine results.

2) Assess the relative merits of various efficiency measures on the "Additional Exercises" Model you created in Chap. 4 using PAT. You may consider some of the same Measures used in Checkpoint Nine but are encouraged to explore other Measures available in the BCL.

# References

ASHRAE (2011) Advanced energy design guide for K-12 school buildings: 50% energy savings

Fleming K, Long N, Swindler A (2012) Building Component Library: an online repository to facilitate building energy model creation. ACEEE summer study on energy efficient buildings, Pacific Grove, CA, August 12–17

https://bcl.nrel.gov/node/39440

https://bcl.nrel.gov/node/39783

https://bcl.nrel.gov/node/82771

https://bcl.nrel.gov/node/83307

https://bcl.nrel.gov/node/83591

https://bcl.nrel.gov/node/83647

https://github.com/NREL/openstudio-standards

https://www.ruby-lang.org/en/

# Chapter 7
# Parametric Analysis with OpenStudio

## 7.1 Introduction to Parametric Analysis

The previous chapter introduced the concept of OpenStudio Measures and how they can be applied individually and in combination to a Model to create and compare different Design Alternatives. While an improvement from modifying models by hand, generating results, and comparing them; the manual analysis workflow is still labor intensive, non-scalable, and will not necessarily yield the best solution for a given problem. In this chapter, we will discuss how OpenStudio enables automated creation and search of large building parameter spaces. We'll also look at how these same approaches may be used to "tune" models of existing buildings to best match measured energy consumption data.

## 7.2 OpenStudio Server

In the previous chapter, we alluded to a "server" that ran on a user's computer to manage multiple simulations in PAT's manual mode. To understand and utilize OpenStudio for parametric analysis, it is important to have a better understanding of what OpenStudio Server is and (to a limited degree) how it works. Figure 7.1 illustrates the basic relationship between PAT, the "OpenStudio Server," and the workers who create and simulate the individual data points.

OpenStudio Server has three primary components:

1. **Web Interface** – A minimal interface that allows for user interaction with analysis projects and the contents of the Results Database,

---

The original version of this chapter was revised. A correction to this chapter can be found at
https://doi.org/10.1007/978-3-319-77809-9_10

**Electronic Supplementary Material:** The online version of this chapter (https://doi.org/10.1007/978-3-319-77809-9_7) contains supplementary material, which is available to authorized users.

© Springer International Publishing AG, part of Springer Nature 2018                    215
L. Brackney et al., *Building Energy Modeling with OpenStudio*,
https://doi.org/10.1007/978-3-319-77809-9_7

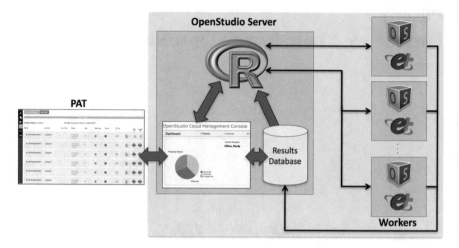

**Fig. 7.1**   PAT, OpenStudio Server, and Workers

2. **Results Database** – A database used to store high level simulation results for a project, and
3. **R** – An open source platform for statistical computing and analysis.[1]

OpenStudio Workers are a fourth component in the overall architecture, separate from but integral to the server. Each worker node is an independent computing instance, configured with OpenStudio, EnergyPlus, and supporting software. When provided with a seed Model, weather file, and one or more Measures; a worker has everything it needs to create, simulate, and post-process results for a given Design Alternative. While workers generate all of the output available in a typical simulation run, file size and storage capacity generally dictate that only a subset of that data be returned to the Results Database for subsequent use. We'll discuss shortly how Reporting Measures, along with PAT's Outputs (⌑) Tab, are used to specify which values are stored.

R does most of the "heavy lifting" for any OpenStudio Server-based analysis, defining individual data points to be simulated. The fundamental difference between the "mini server" used locally for manual analyses and a full OpenStudio Server implementation used in algorithmic mode is the inclusion of R and supporting files. The manner in which R defines data points, reviews results, prescribes additional points, etc. varies based on the algorithm chosen, and is discussed in the following section.

Lastly, OpenStudio Server manages an API that translates PAT projects into problem formulations for R, communication with worker nodes, queuing of simulation jobs, and communication of results back to PAT. Additional details of OpenStudio Server are presented in Chap. 9.

---

[1] https://www.r-project.org/.

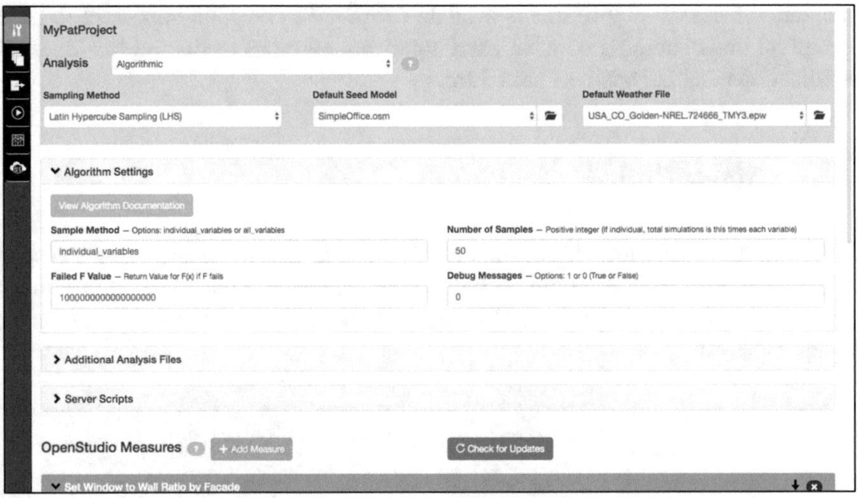

**Fig. 7.2**  PAT analysis Tab in algorithmic mode

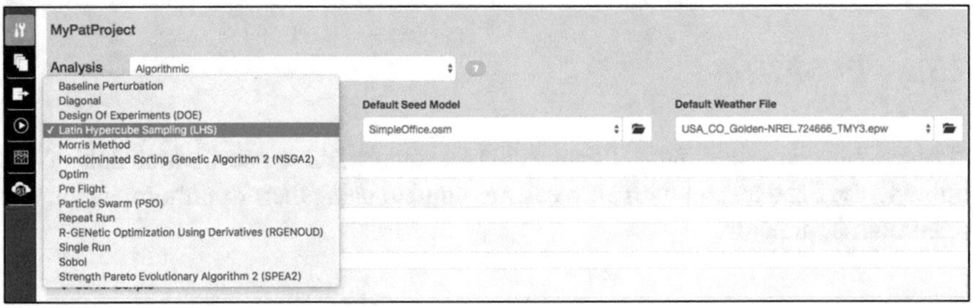

**Fig. 7.3**  Selecting an algorithm in PAT

## 7.3  Algorithms in PAT

PAT has been designed to enable large-scale exploration of design spaces using a range of sampling, optimization, and machine learning algorithms. Switching from "Manual" to "Algorithmic" in the Analysis selection field of the Analysis (⬛) Tab modifies PAT's interface and functionality in a number of ways. Immediately obvious are the additions of "sampling method" as a selection field near the top of the window along with new sections in the Tab for algorithm settings, analysis files, and server scripts (Fig. 7.2).

In this example, Latin Hypercube Sampling (LHS) is selected as the algorithm that will be used to guide exploration of the design space. Immediately below the sampling method field are collapsible sections denoted by a ▸ Button for algorithm settings, supplementary analysis files, and server scripts. A ▭ View Algorithm Documentation ▭ Button provides a link to detailed documentation for each algorithm and its settings.

Figure 7.3 shows the contents of the sampling method choice list and the range of algorithms that OpenStudio server currently supports. While complete documentation for most algorithms is available via ▭▭▭▭▭, it's important to have a general understanding of what each algorithm attempts to do and the kinds of applications that each is best suited for.

### 7.3.1  Single Run

Selecting the Single Run algorithm creates a single data point including the seed Model and each Measure applied with its static/default value. This algorithm is primarily used for testing the seed Model and Measures on a server prior to running a larger analysis.

> **Tip:** This algorithm is so-named because it is intended as a "pre-flight" check of the seed model and measures prior to performing any larger analysis. Pre-flight checks are strongly recommended prior to committing significant computing resources for sampling and optimization problems.

### 7.3.2  Pre-Flight

"Pre-Flight" takes the Single Run algorithm a step further, creating three data points utilizing the Seed Model with all measures applied using their minimum, mean, and maximum arguments.

### 7.3.3  Repeat Run

This algorithm replicates the data point defined by the Single Run algorithm a specified number of times. It is primarily used for software development and testing of the OpenStudio Server itself, and is not of general interest.

### 7.3.4  Baseline Perturbation

This is an experimental algorithm and is not recommended for general use at the time of writing.

### 7.3.5  Diagonal

This is an experimental algorithm and is not recommended for general use at the time of writing.

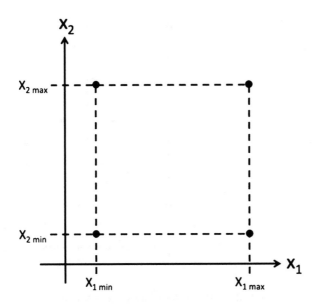

**Fig. 7.4** A DOE with two independent variables and two sampling levels

## 7.3.6  Design of Experiments (DOE)

Design Of Experiments or "DOE" is a sampling method that varies each independent variable separately. The number of samples (sometimes referred to as levels) in a DOE problem determines how many times the effect of each independent variable will be evaluated. So a two level DOE with two independent variables would sample the design space shown in Fig. 7.4 resulting in four total data points.

The "full factorial" DOE implemented in OpenStudio Server generates a total of $m^n$ data points for a problem with "n" independent variables and "m" levels,. With problems of any significant complexity, the total number of simulations to be performed in a DOE quickly grows out of control. For this reason, mathematicians have devised alternate sampling methods that attempt to reduce the number of samples without greatly compromising the quality of an analysis.

## 7.3.7  Latin Hypercube Sampling (LHS)

Latin Hypercube Sampling,[2] or "LHS," attempts to generate a sparser set of samples than DOE using pseudo-random distributions of independent variables. In LHS, independent variable ranges are divided into segments containing equally probable

---

[2] McKay et al. (1979).

**Fig. 7.5** LHS concept
with two independent
variables and four samples

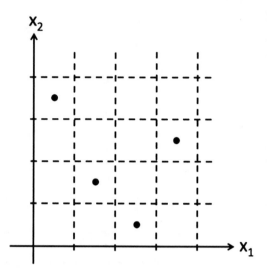

intervals. The concept of a "Latin Square," a square array in which array elements appear exactly once in each row and column, is applied to those independent ranges to define samples. A simple example involving two independent variables and four samples is shown in Fig. 7.5.

Note that number of samples dictates the number of intervals applied to each independent variable and consequently, the total number of data points that will be defined. For a fixed number of samples, as the number of independent variables increases, the multi-dimensional "grid" discretizing the sample space becomes coarser and coarser, resulting in fewer data points in each dimension, and a rougher approximation of the variable's distribution. We will revisit the concept of independent variable distributions in Sect. 7.4, but for the moment think of LHS as a reasonably efficient means of assessing how different design parameters (Measure arguments) affect building performance.

## 7.3.8  Morris Method

The previous algorithms generate data points irrespective of their relative performance in terms of energy use intensity (EUI), cost, or whatever other combination of criteria might be of interest. The Morris Method and optimization algorithms described in Sects. 7.3.9 through 7.3.13 all rely on knowing something about how well any given data point achieves a performance objective. Abstractly, we define a performance objective function "F" in terms of our independent variables, the vector **x**. We will discuss exactly how we define F(**x**) using PAT in Sect. 7.5.1, but for the moment imagine that algorithms have a means of quantifying how well any given data point meets an objective like EUI.

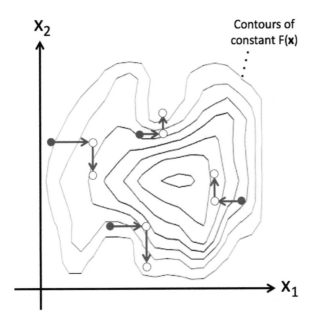

**Fig. 7.6** Morris method concept with initial and subsequent data points for r = 4

The Morris Method[3] is used to assess how sensitive the objective function is to changes in the independent variables. For example if window-to-wall ratio and insulation R value were both available as Measure arguments in a PAT analysis, the algorithm can be used to inform which of those variables had a more significant impact on EUI for a particular building. Unlike DOE or LHS, which identify all samples (and associated input argument values) a priori, Morris method starts by "randomly" selecting a set of samples within the available input ranges, simulating those data points, and evaluating the objective function for each. Input arguments are modified, one at a time, creating new data points, simulations, and objective functions. A new starting point is randomly selected and the entire process is repeated a number of times. If r is the number of times and n is the number of independent variables, this produces a total of r * (n + 1) data points (Fig. 7.6).

## 7.3.9   Sobol Method

The Sobol[4] Method is an alternative to Morris for identifying parametric sensitivity for a problem. Sobol decomposes the variance of outputs into fractions that are attributed to inputs. Sobol typically requires more point evaluations than Morris, providing better coverage of the parameter space at the expense of computation time.

---

[3] Morris (1991).
[4] Sobol (2001).

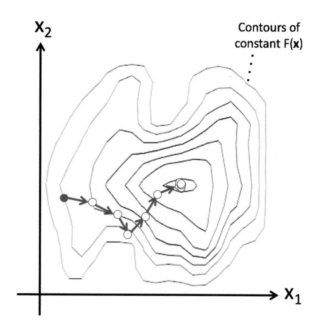

**Fig. 7.7**  Optim conceptual progression to optimal solution

### 7.3.10  Fourier Amplitude Sensitivity Test (FAST99) Method

The Fourier Amplitude Sensitivity Test (FAST99)[5] is a third method of estimating parametric sensitivity based on a variation of the Sobol Method. It improves (computationally) on Sobol, and generally falls somewhere between Morris and Sobol in terms of computational performance and accuracy.

### 7.3.11  Optim

Optim is the simplest goal-seeking algorithm supported by OpenStudio Server. It is based largely on a Nelder-Mead Simplex method[6] that moves in single directions (linear combinations of **x**) until no further improvement in F(**x**) is achieved. Figure 7.7 shows a conceptual progression of a Simplex approach for two independent variables and a well-behaved performance surface. Because optim calculates approximate derivatives for the performance index, it is <u>not</u> recommended to apply it to problems with non-differentiable surfaces. This includes design problems with discrete independent variables (e.g. wall insulation values of R20, R30, R40, etc.)

---

[5] Saltelli et al. (1999).
[6] Nelder and Mead (1965).

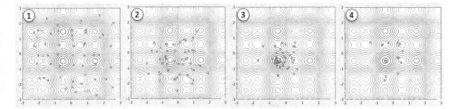

**Fig. 7.8** Example particle swarm optimization progression (Figure adapted from Ephramac, 2017)

Since a great many building optimization problems of interest fall into this category, optim should always be used with an abundance of caution.

## 7.3.12   Particle Swarm Optimization (PSO)

Particle Swarm Optimization was originally intended to simulate social behavior, and loosely modeled the behavior of flocks of birds and schools of fish.[7] In this algorithm a set of particles (data points) migrate around the design space based on an equation that governs their relative velocities, proximity to local optimae, and proximity to the most optimal solution observed by the swarm. Over time, the swarm generally drifts towards the best known solution, but individual particles are granted a degree of autonomy to explore local optimae that may yield better solutions (Fig. 7.8). This potentially allows the algorithm to identify a global optimum when performance surfaces contain multiple local minimae. Because the algorithm does not require the performance surface be differentiable, it is better suited to address a broader class of optimization problems than "optim."

## 7.3.13   Nondominated Sorting Genetic Algorithm 2 (NSGA2)

The last three optimizers discussed in this section are based (at least in part) around a class of solvers referred to as "genetic algorithms," which attempt to digitally mimic the concept of natural selection.[8,9] In a genetic algorithm, the independent variables are typically discretized or "encoded" to construct a population of solutions. Figure 7.9 contains examples for three data points with roof and wall insulation, lighting, and window-wall-ratio as independent variables; effectively identifying the "DNA" of a particular building. In the case of variables that are already discrete (e.g. insulation values of R20, R30, R40, etc.) no encoding is necessary – this enables genetic methods to address building design problems that other algorithms cannot.

---

[7] Kennedy and Eberhart (1995).

[8] Barricelli (1957).

[9] Fraser (1957).

**Fig. 7.9** Genetic
algorithm "DNA" for three
building data points

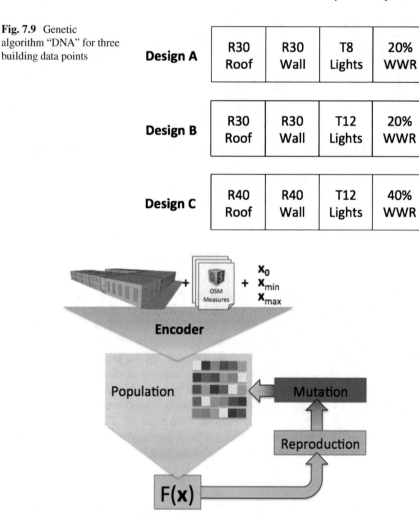

| **Design A** | R30 Roof | R30 Wall | T8 Lights | 20% WWR |

| **Design B** | R30 Roof | R30 Wall | T12 Lights | 20% WWR |

| **Design C** | R40 Roof | R40 Wall | T12 Lights | 40% WWR |

**Fig. 7.10** Genetic algorithm applied to a building optimization problem

Once a population has been established, the data points are simulated and $F(\mathbf{x})$ is evaluated. The next generation's "offspring" are created by combining data points (parents) that performed well. Poorly performing data points are removed from the "gene pool," and the process repeats. Genetic algorithms may also introduce the concept of "mutation," random modifications to an offspring's DNA that enable the algorithm to identify better solutions by chance. This iterative process is illustrated in Fig. 7.10.

Unlike other optimizers, NSGA2 does not require a scalar objective function. As generations progress from an initially random population, a locus of points, $F(\mathbf{x})$, is generated that represent the best identified solutions that have been observed. This locus is called a "Pareto front," and represents a collection of

**Fig. 7.11** Multi-variate optimization of heating and cooling energy use with Pareto Front

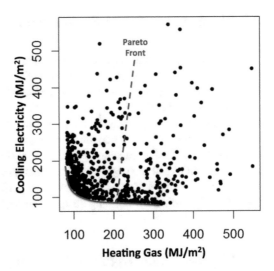

solutions that are "good" in some sense. Figure 7.11 illustrates the results of a multi-variate optimization that seeks to minimize heating and cooling energy use. Which solution is best? The answer to that question is – it depends. For this problem do we value cooling electricity use more than heating gas consumption? The Pareto Front illustrates the tradeoff between these competing goals, and the "best" solution likely exists somewhere on that curve. One solution might be to apply the associated energy costs as weighting factors to effectively turn the problem into a univariate optimization problem, but as we'll see in subsequent examples, a singularly optimal solution is rarely that straightforward.

In the case of the NSGA2 algorithm, the relative distances between data points and the Pareto Front are used to sort or rank solutions for pairing and reproduction.[10] The algorithm also gives preference to solutions which are not clumped together near other solutions, promoting their genetic characteristics into subsequent generations. As with other genetic algorithms, the process proceeds until a specific number of generations has been reached, or an optimization convergence criteria has been met.

### 7.3.14  Genetic Algorithm with Island Evolution (GAISL)

GAISL is another genetic algorithm variant that partitions populations into isolated "islands" for reproduction and mutation.[11] Occasionally a data point will "migrate" to another island, allowing subpopulations to share genetic material.

---

[10] Deb et al. (2002).
[11] Belding (1995).

### 7.3.15   Strength Pareto Evolutionary Algorithm 2 (SPEA2)

As its name might suggest, SPEA2 is also a multi-objective optimization algorithm that makes use of Pareto Front information to inform how the algorithm progresses. SPEA2 is also a genetic algorithm and shares many similarities to the NSGA2 approach. SPEA2 differs in how it utilizes Pareto information to cluster data points and match pairs for propogation into subsequent generations.[12] Comparison of SPEA2 and NSGA2 for a specific power system dispatch optimization problem suggests that SPEA2 is capable of finding slightly better/more diverse sets of solutions than NSGA2 at the expense of increased computation time due to its computational complexity.[13] The authors have not performed extensive comparisons of these two algorithms for building design optimization problems.

### 7.3.16   R-GENetic Optimization Using Derivatives (RGENOUD)

RGENOUD, is a hybrid solver that combines the strengths of a genetic algorithm (discrete variable capability, robustness in finding global optimum, etc.) with the potential speed of gradient-based solvers like optim.[14] In the case of RGENOUD, derivative estimates inform propogation between generations to quickly climb hills or decend valleys of the performance surface without significantly compromising the algorithms stability and likelyhood of identifying a global minima. The authors regularly use RGENOUD for building optimization problems, and generally that it offers a reasonable trade-off between computation speed and performance.

## 7.4   Working with Measure Variables and Arguments

Algorithmic mode significantly alters the Measure section of the Analysis (▣) Tab. The concept of manual mode's Measure "option" no longer applies, and variable checkboxes now become selection fields that allow the user to specify the nature of Measure inputs. Measure inputs may be set to one of four types in PAT:

1. **Argument** – Fixes the quantity as a static value for the analysis.
2. **Continuous** – Assigns a continuous probability density function to a variable.
3. **Discrete** – Allows the user to specify a specific set of distinct values with associated weights.
4. **Pivot** – Forces the entire analysis to be repeated against each specified (discrete) value.

---

[12] Zitzler and Thiele (1998).

[13] King et al. (2010).

[14] Mebane and Sekhon (2011).

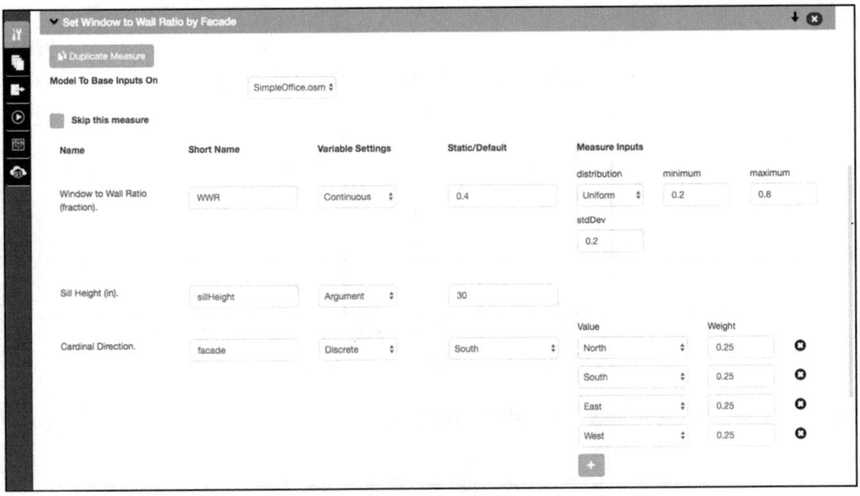

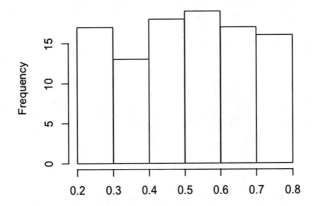

**Fig. 7.12**  Assigning multiple variable types to a Measure's inputs

**Fig. 7.13**  Uniform distribution for window to wall ratio

By default, all Measure inputs are considered arguments unless explicitly changed to another type. This allows the user to vary, for example, only the window-to-wall ratio in the Set Window to Wall Ratio by Façade Measure. Multiple inputs in a single Measure may vary, and they need not all be defined as the same variable type. Figure 7.12 provides a good example involving the Window to Wall Ratio by Façade Measure. In this case, the window to wall ratio is a continuous variable, sill height is a constant, and the cardinal direction of the windows changed by the Measure will vary according to a discrete distribution.

Figure 7.13 shows the distribution of window to wall ratio values that the Measure might apply to the seed Model over 100 simulations. As the number of simulations increases, the histogram would become flatter, better approximating the uniform distribution requested.

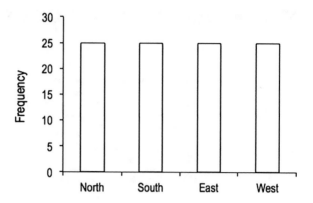

**Fig. 7.14**  Discrete distribution for cardinal direction

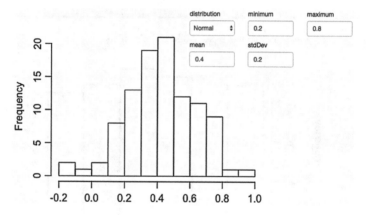

**Fig. 7.15**  Normal distribution for window to wall ratio

A discrete distribution for the side of the building the Measure will alter windows on is shown in Fig. 7.14 for a population of 100 data points. The discrete option is appropriate here because the cardinal direction argument in the Measure is a choice list comprised of North, South, East, and West. Additional options may be added by clicking the ■ Button and removed with the ○ Buttons next to each value/weight pair. The weight values are normalized to determine how frequently a value might appear in a population. In this case, each of the options has a 25% chance of occurring.

PAT supports three additional types of distributions for sampling problems involving continuous variables: normal, log normal, and triangular that are commonly used in statistical analysis and modeling. Most readers will likely be familiar with the normal distribution shown in Fig. 7.15. Changing the distribution type in PAT adds a mean input field. In this case, we've requested that the algorithm produce a distribution of window to wall ratio values with a mean of 0.4 and standard deviation of 0.2. The histogram shows a possible distribution of Measure inputs over 100 simulations.

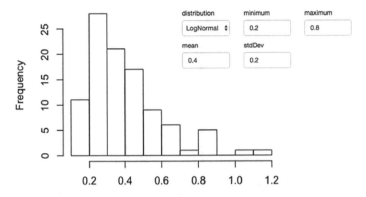

**Fig. 7.16**  LogNormal distribution for window to wall ratio

Figure 7.16 illustrates distribution of window to wall ratios one might expect when a log normal distribution is applied to a population of 100 simulations.

Note in both the normal and log normal cases that the minimum and maximum entries in PAT do <u>not</u> limit the possible input values that might be generated by a sampling algorithm.[15] In the case of the normal distribution there are extreme results in the distribution's tail with negative window to wall ratios. In the log normal case, the Measure would be asked to modify the seed Model to produce window to wall ratios exceeding 1.0, also a physical impossibility. Well-written Measures will "trap" such situations and limit inputs to reasonable values, reporting a warning in the process. We'll discuss best practices for writing Measures in Chap. 9, but is it safe to assume that all Measures will be sufficiently robust against nonsensical inputs? Because of this possibility, the authors recommend using the triangular distribution in lieu of others.

Figure 7.17 illustrates a potential triangular distribution of window to wall ratio over 100 samples. Results are strictly constrained by the minimum and maximum values, with the peak of the triangle occurring near the mean (or more accurately the mode) of the triangular distribution. Triangular distributions can be used to roughly approximate both normal and log normal distributions, while ensuring that Measure inputs and resulting data points are reasonable.

That final variable type supported by PAT is a "pivot." At first glance, Pivots might appear to behave like discrete variables, as the user can enter values with the ◼ Button and remove them with the ○ Buttons. However, one complete sampling problem is performed for each value of a pivot variable. An analysis problem with 100 samples and 3 pivots would create a total of 300 data points to be simulated. Why might you use a pivot variable? One common use for them is not in performing an analysis of a single building, rather studying how efficiency measures might apply to many types of buildings. Figure 7.18 illustrates just this situation.

---

[15] While samplingalgorithms (e.g. LHS) will not respect prescribed maximum and minimum values, they are used to constrain solutions generated by PAT's optimizers.

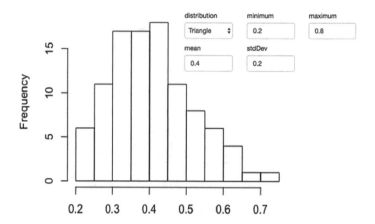

**Fig. 7.17** Triangle distribution for window to wall ratio

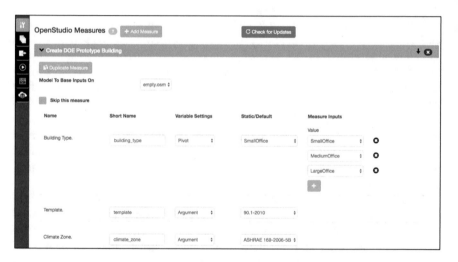

**Fig. 7.18** "Pivoting" an analysis on building type

In this example, the "Create DOE Prototype Building" Measure and OpenStudio Standards Gem introduced in Chap. 6 are used to create an entire building Model from scratch. Note that the seed Model in this project is called "empty.osm," which is literally an empty Model. The building type input for the Measure is set as a pivot with Small, Medium, and Large Office as values. Any of the Measure's 13 other building types including schools, hospitals, hotels, and retail could also be added to expand the study. The building template and climate zone inputs are left as arguments for ASHRAE 90.1-2010 and climate zone 5B respectively. Combining this with the "Set Window to Wall Ratio by Façade" Measure along with an appropriate sampling algorithm allows us to investigate the effect of window variation on multiple building types. This is an excellent example of the kind of flexibility that OpenStudio, Measures, and PAT provide.

## 7.5   Other Changes to PAT in Algorithmic Mode

In addition to the option to select and configure algorithms in the Analysis (▥) Tab, a number of other changes manifest throughout PAT's tabs – some obvious and some subtle. For example, since an algorithm is used to specify all Design Alternatives, there is no use for PAT's Design Alternatives (▢) Tab as shown in Fig. 7.19.

### 7.5.1   The Outputs Tab in Algorithmic Mode

Analysis via algorithm tends to produce very large data sets. As such, the methods for visualizing results and teasing out valuable insights differ from the simple table views available in manual mode. In addition, some algorithms (e.g. optimizers) are goal seeking, and require specification of performance metrics to function properly. PAT's Outputs (▣) Tab, shown in Fig. 7.20, is designed to specify key simulation outputs for use in post-processing large sets of simulation results or as a means of guiding optimizers. At a minimum, the Outputs (▣) Tab needs the OpenStudio Results Measure to be available to provide a nominal set of outputs, although additional reporting Measures may be used to add more outputs.

Pressing the ▭Select Outputs Button brings up a dialog of all outputs specified by the reporting Measure (Fig. 7.21). The user may select all of the outputs, or only individual outputs via check boxes.

**Fig. 7.19**   PAT's Design Alternatives Tab in algorithmic mode

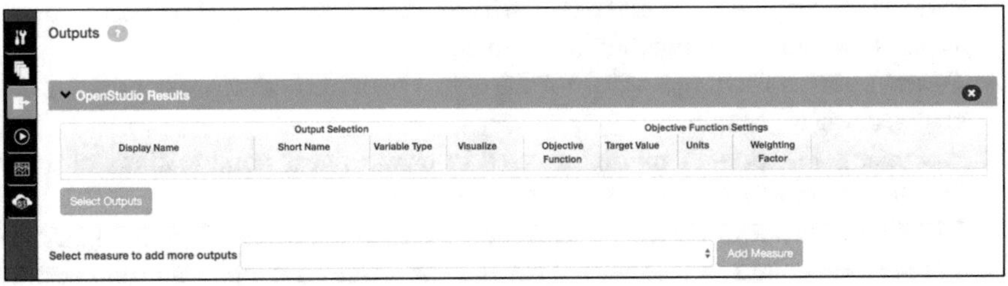

**Fig. 7.20**   PAT's Outputs Tab in algorithmic mode

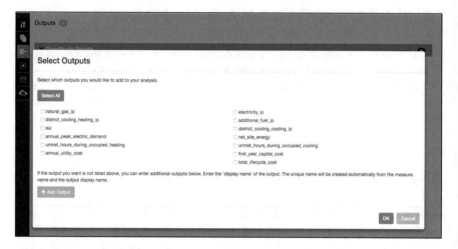

**Fig. 7.21** Outputs available from the OpenStudio Results Measure

**Fig. 7.22** Configuring selected outputs

Once selected, outputs appear in the Outputs (▣) Tab along with additional fields (Fig. 7.22). The "visualize" field should be set to true for any variables the user wishes to preserve from worker node simulations for use in visualizations or subsequent analysis.

The remaining columns in the Outputs (▣) Tab are used to specify objective function, F(**x**). They include:

1. **Objective Function** – Should be set to true for any variables that will be included as part of an objective function (an element in vector **x**).
2. **Target Value** – The target value for the output variable that will be sought by the goal-seeking algorithm.
3. **Weighting Factor** – A multiplier used to create linear combinations of error functions in multi-objective optimization problems.
4. **Objective Function Groups** – This argument only appears when an algorithm that performs multi-objective analysis is in use. Use an integer to group performance indices or assigning each index to its own group number.

It is important to note that many optimization algorithms require scalar performance metrics. Multi-objective optimization is frequently achieved through linear combination (weighted average) of separate performance indices as shown in the following equation:

$$F(x) = w_1 \sqrt{(f_1 - t_1)^2} + w_2 \sqrt{(f_2 - t_2)^2} + w_n \sqrt{(f_n - t_n)^2}$$

where:

$w_i$ is the $i$th weighting factor,
$f_i$ is a performance metric computed by simulating a data point, which is a function of $\mathbf{x}$, and
$t_i$ is a target value associated with the $i$th performance metric.

The $\sqrt{(f_i - t_i)^2}$ terms in the equation represents an "$L^2$ norm" of the error between metric and target. Although some algorithms can use other norms, the L2 norm is commonly used by goal-seeking algorithms that require a positive definite performance index.

## 7.5.2   The Run Tab in Algorithmic Mode

The size of algorithm-based analysis problems generally exceeds the computational capability of a personal computer. PAT has been designed to run these sorts of analyses in the cloud or on dedicated servers. While PAT's "mini server" still launches in the background, the application does not support algorithms using local computing resources. For that reason, PAT's Run (▣) Tab will initially look like Fig. 7.23 when first selected in Algorithmic Mode.

Selecting "Run on Cloud" as the Run option changes the Run (▣) Tab significantly (Fig. 7.24).

Run on Cloud produces a dialog that enables the user to specify either an "Existing Remote Server" or Amazon Cloud. Provisioning an existing, dedicated

**Fig. 7.23**   PAT's Run Tab in algorithmic mode

**Fig. 7.24**  PAT's Run Tab with Run on Cloud selected

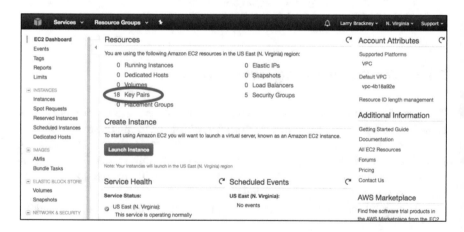

**Fig. 7.25**  AWS console with secret key option highlighted

server with OpenStudio is beyond the scope of this text. Interested users are referred to https://github.com/NREL/OpenStudio-server for further guidance on the topic.

Amazon Web Service (AWS) has been pre-provisioned with Amazon Machine Images (AMIs) for OpenStudio Server that enables users to leverage large-scale computing with minimal effort and cost. New users will need to establish an account at https://aws.amazon.com. The AWS web console enables tracking and management of computing resources, security credentials, billing details, etc. Figure 7.25 depicts the AWS console with a link to its "Key Pair" management tool highlighted. Establishing a key pair is the first step in linking OpenStudio to AWS. Clicking on this link takes the user to a management screen with a ▭ Create Key Pair Button allowing creation of a new set of credentials. Follow the instructions on AWS and make note of the new access and secret keys before proceeding.

The following steps begin the process of setting up PAT to use AWS as a remote computing cluster for large-scale analysis:

1. In the PAT Run (▣) Tab, select "Run on Cloud" with "Amazon Cloud" as the server type.
2. Press the [New Cluster] Button and choose a cluster name to store your AWS cluster settings.
3. Press the [New] Button in the AWS Credentials area and follow the prompts to enter the access and secret key information created from the AWS console.

The cluster name is used to save PAT's settings for AWS including the number of workers, size of compute nodes, etc. Multiple setting sets may be stored for use as an analysis is scaled up. Like the cluster name, the AWS credential name is used by PAT to help the user manage access credential configurations, and multiple sets may be stored to assign analyses to different AWS accounts.

A field labeled "AMI Name" specifies the version of OpenStudio that the server and worker nodes will use. This list may include a large number of major and minor OpenStudio versions. It is recommended that you use a major release (one ending in a .0) that corresponds with the version of OpenStudio installed on your computer. The specific version of OpenStudio and the Standards Gem included in the AMI are listed next to the AMI Name field.

The Server and Worker Instance Type fields allow the user to configure the processing capability and storage size of the primary server and workers used to perform an analysis. The user is referred to AWS documentation related to server and worker options, but PAT provides a brief description of the node configuration and approximate cost/hour next to the server choice field. A PAT analysis requires a server and a minimum of one worker. Figure 7.26 illustrates a cluster configured to run an analysis with an "m3.2xlarge" server and 2 m3.2xlarge workers at a total cost of approximately $1.68[16] per hour. Each machine has 8 CPUs so a total of 16 workers along with a few of the server's spare nodes are available for simulation.

Once configured and saved with the [Save Cluster Settings] Button, the user should press the [Start] Button next to the Cluster Status indicator on the right side of the UI. A dialog (Fig. 7.27) explaining the user's responsibility to monitor and manage AWS computing resources appears and must be acknowledged before proceeding.

At this point, the server and worker provisioning process begins as shown in Fig. 7.28. A startup dialog states that the process may take many minutes.

Clicking on the [View AWS Console] Button will open the AWS console in a web browser, allowing the user to monitor the startup process as shown in Figs. 7.29 and 7.30.

Once the server is running, the Server Status indicator will change from ● to ●. Two new Buttons also appear in the Run (▣) Tab: [Run Entire Workflow] and [View Server]. Clicking the [Run Entire Workflow] Button starts the analysis on the server. As in Manual mode, progress may be monitored within PAT as shown in Fig. 7.31. The [View Server] Button functionality will be discussed in the next section.

---

[16]Amazon routinely updates its pricing structure. Prices listed in PAT are only estimates, and the user is ultimately responsible for knowing what costs may be incurred by an analysis.

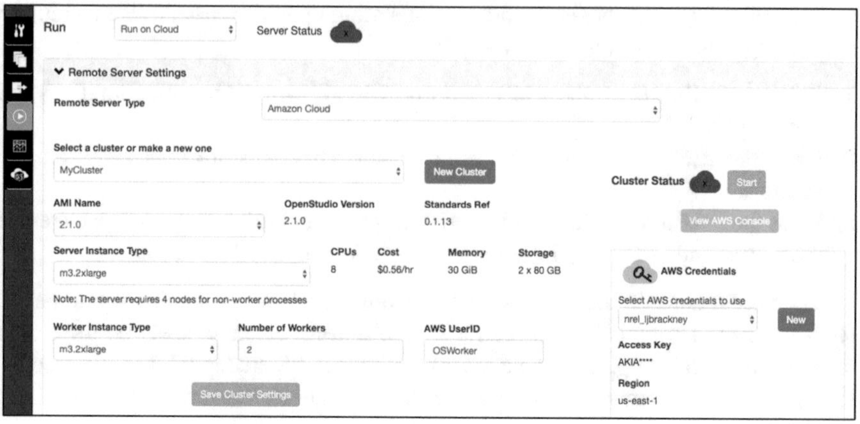

**Fig. 7.26**  Server and worker node configuration in PAT's Run Tab

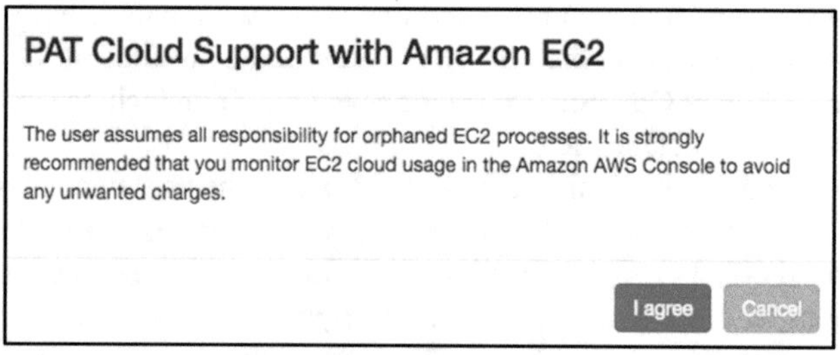

**Fig. 7.27**  PAT AWS responsibility dialog

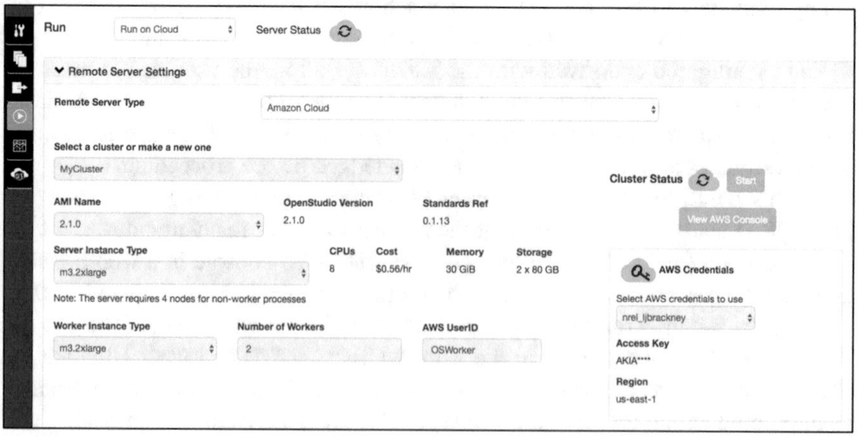

**Fig. 7.28**  AWS cluster starting up in PAT

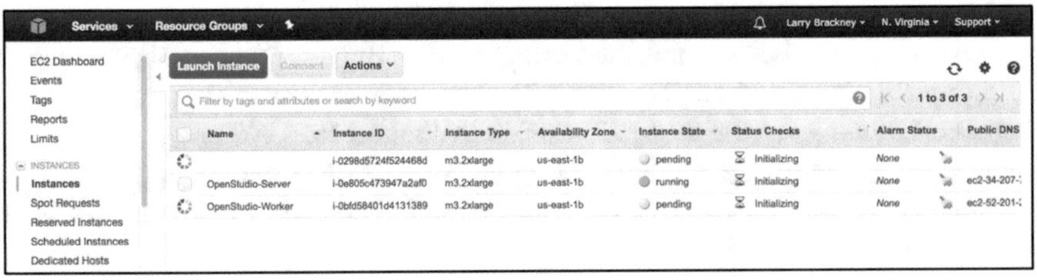

**Fig. 7.29** AWS cluster starting up in the AWS console

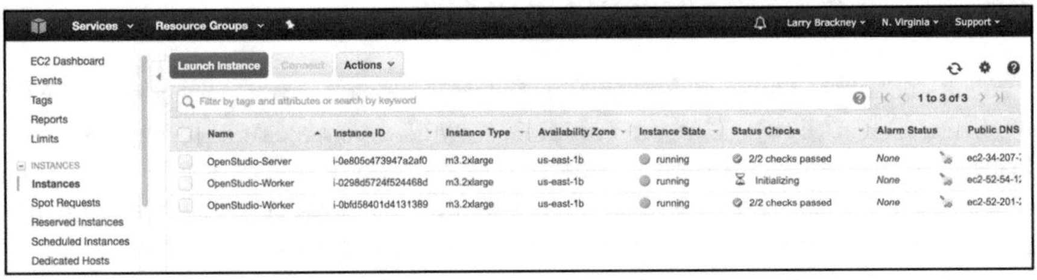

**Fig. 7.30** AWS cluster nearly ready for use in the AWS console

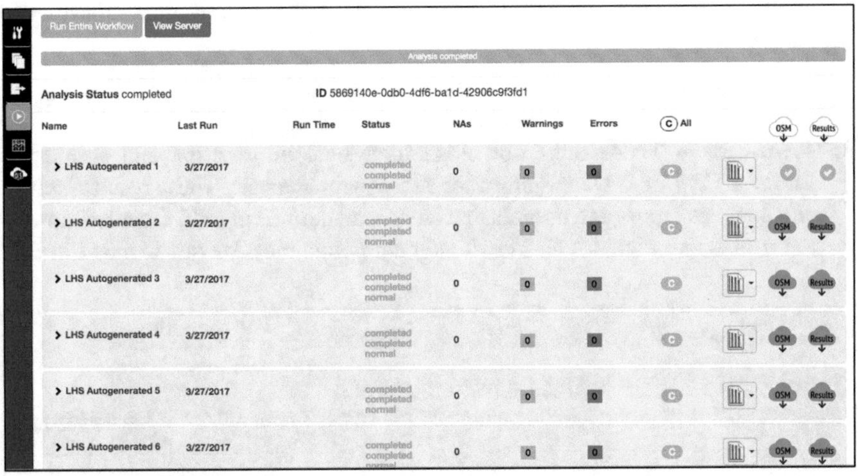

**Fig. 7.31** Cloud run completed

One notable difference between running in manual mode with local computing resources and on the server is that detailed simulations are not automatically downloaded from the cloud. These files can be quite large. Clicking on the ✚ or ✚ Buttons next to a data point will download the OpenStudio Model or completed data point zip file. The cloud Buttons appear as a ◎ when a download has occurred. Data point files are lost when the server shuts down, so it is important to download results that may be of particular interest while the server is running.

When you are finished using AWS, it is important to shut the cluster down properly. This can be accomplished using the ▣ Button. Quitting PAT will also bring up a dialog providing the user with the choice of terminating the cluster or leaving it running for the next PAT session to connect to. <u>AWS nodes continue to incur costs to the user regardless of whether simulations are running or not.</u> It is recommended that the user quickly check the AWS console when finishing work to make sure all instances are terminated as expected.

## 7.6  Working with OpenStudio Server

The OpenStudio Analysis Server (▣) Tab is of much greater importance with algorithmic workflows. This same content may also be accessed through any web browser by clicking the ▣ Button in the Run (▣) Tab. The top-level view of OpenStudio server, shown in Fig. 7.32, provides a summary of completed or in-progress projects and analyses along with navigation options.

A single OpenStudio Server analysis (a specific run of a project) is shown in Fig. 7.33. This view provides a high-level summary of the project, links to more detail about the analysis, and status updates for all data points that have been completed, queued for simulation, or are in process. A few important links on this page include:

1. **Project Log** – Near the bottom of the Analysis Information box is a view log file link, which can be helpful in debugging failed analyses.
2. **Downloads** – These links download high-level information and simulation results as CSV or R Data Frames for subsequent analysis. These results include only analysis inputs and outputs that have been defined in PAT. Detailed simulation results associated with individual data points must be downloaded in PAT

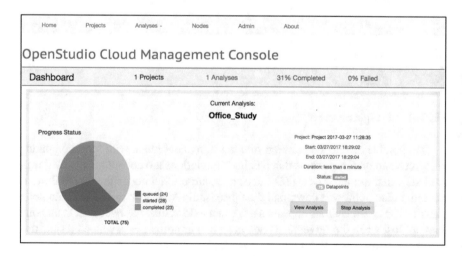

**Fig. 7.32**  OpenStudio Cloud Management Console front page

Home        Projects        Analyses ▾        Nodes        Admin        About

## OpenStudio Cloud Management Console

MyPatProject

| Analysis Information | | Data and Visualizations ④ | Downloads ② |
|---|---|---|---|
| Project | Project 2017-03-27 08:58:18 | Measures (3) | Seed Zip File |
| Type | lhs | Variables (3 Perturbable) | CSV (Metadata) |
| Status | completed | Analysis Data | CSV (Results) |
| Status Message | | Parallel Coordinates Plot | R Data Frame (Metadata) |
| Start Time | 03-27-2017 14:32:26 UTC | Scatter Plot | R Data Frame (Results) |
| End Time | 03-27-2017 14:32:30 UTC | Interactive XY Plot | |
| Duration | less than a minute | | |
| View | JSON \| Log ① | | |

### Simulations (104 / 104 )

All    Completed    Started    Queued    Initialized

**All Simulations**                                                    [Search]

View All

| Name | Status | Status Message | Start Time (UTC) | End Time (UTC) | Delta Time (s) | View | Action |
|---|---|---|---|---|---|---|---|
| LHS Autogenerated 1 | completed | completed normal | 03-27-2017 14:32:30 | 03-27-2017 14:33:24 | 54.41 | View \| JSON \| Zip File | Destroy |
| LHS Autogenerated 2 | completed | completed normal | 03-27-2017 14:32:30 | 03-27-2017 14:33:24 | 54.04 | View \| JSON \| Zip File | Destroy |
| LHS Autogenerated 3 | completed | completed normal | 03-27-2017 14:32:30 | 03-27-2017 14:33:24 | 53.55 | View \| JSON \| Zip File | Destroy |
| LHS Autogenerated 4 | completed | completed normal | 03-27-2017 14:32:31 | 03-27-2017 14:33:24 | 53.58 | View \| JSON \| Zip File | Destroy |
| LHS Autogenerated 5 | completed | completed normal | 03-27-2017 14:32:31 | 03-27-2017 14:33:29 | 57.72 | View \| JSON \| Zip File | Destroy |

③

**Fig. 7.33**  OpenStudio Cloud Management Console view of a project

or via individual data point web links. Additional downloads may be available here upon completion of an analysis depending upon the algorithm being used.

3. **Data Points** – The bottom of the analysis page includes a snapshot of all data points along with their status, run times, and data point-specific links including a zip file containing models and simulation results.

4. **Data and Visualizations** – A number of useful project summaries and interactive visualization tools are built into OpenStudio server.

The variables link near the top center of the analysis web page provides a concise summary of variables and arguments utilized by the analysis algorithm. While any applicable Measures are summarized in an adjacent page, this section provides more detail about how data points have been generated. In the example below, building type is used as a pivot variable as shown in Fig. 7.34. A five level "Design of Experiments" with lighting power density reduction percentage and window to wall ratio variables complete the study space.

Parallel coordinate plots are one of the useful visualization tools built into OpenStudio Server (Fig. 7.35). They provide an interactive means of exploring large data sets and teasing out valuable insights. The plotting tool enables the user to select inputs and outputs that have been pre-defined in PAT's Analysis and

Home          Projects          Analyses -          Nodes          Admin          About

## OpenStudio Cloud Management Console

### Variables

Modify Variables | Download Variables as R Data Frame | Download Variables as CSV | JSON

### Pivot Variables

| Display Name | Name | Number of Samples | Thumbnail |
|---|---|---|---|
| Building Type. | create_doe_prototype_building.building_type | 3 | No image |

### Variables Variables

| Display Name | Name | Number of Samples | Thumbnail |
|---|---|---|---|
| Lighting Power Reduction (%). | reduce_lighting_loads_by_percentage.lighting_power_reduction_percent | 5 | No image |
| Window to Wall Ratio (fraction). | set_window_to_wall_ratio_by_facade.wwr | 5 | No image |

### Output Variables

| Display Name | Name | Taxonomy ID | Units | Objective Function Index |
|---|---|---|---|---|
| annual_peak_electric_demand | OpenStudioResults.annual_peak_electric_demand | | | |
| eui | OpenStudioResults.eui | | | |

### Other Variables/Arguments

| Display Name | Name | Units |
|---|---|---|
| Air Loops Detail | air_loops_detail_section | |
| Annual Overview | annual_overview_section | |
| Model Summary | building_summary_section | |
| Climate Zone. | climate_zone | |
| Cash Flow | cost_summary_section | |
| Demolition Costs Occur During Initial Construction? | demo_cost_initial_const | |
| Increase in Demolition Costs for Lighting per Floor Area (%). | demolition_cost | |
| Select an Electricity Tariff. | elec_tar | |
| Envelope | envelope_section_section | |
| Climate File (NECB nch) | epw_file | |

**Fig. 7.34**  OpenStudio Cloud Management Console project variable summary

Outputs Tabs. Checkboxes turn inputs on and off, and the individual plot axes can be re-ordered via drag and drop. In this example, LHS is used to sample the parameter space for lighting power density, window to wall ratio, and the facade on which windows are placed. Energy Use Intensity (EUI) is selected as the output of interest.

Each blue-segmented line in the parallel coordinate plot represents a single data point. Segments connect key simulation inputs with the resulting outputs. In the above example: lighting power density reduction, window to wall ratio, and window cardinal direction inputs appear on the left side of the chart with EUI as the output of interest on the far right. Figure 7.36 is an annotated parallel coordinate plot highlighting a single data point in a net zero building study performed with PAT.

More than a static display, the user can filter the data points based on input or output ranges to zero in on simulations with certain characteristics. For example, Fig. 7.37 shows how filtering is used in the net zero study to highlight those data points that are net zero or net positive energy producers. We'll make use of this capability in subsequent sections to gain insight into modeled performance.

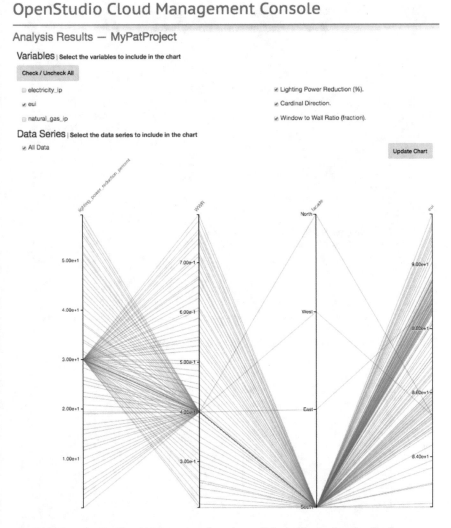

**Fig. 7.35**  OpenStudio Cloud Management Console parallel coordinate plot visualization

## 7.7   Checkpoint Ten: Sampling Problems and Uncertainty Analysis

The next two checkpoint exercises assume you have created an AWS account to use EC-2 resources, or that your organization has provisioned a dedicated OpenStudio Server for you to use. The most intensive of these exercises can be analyzed in 10–20 min using 40 or more worker nodes. This first exercise focuses on establishing the analysis setup that we will use in Checkpoint Eleven to solve a problem that building scientists and practitioners are often faced with – Model

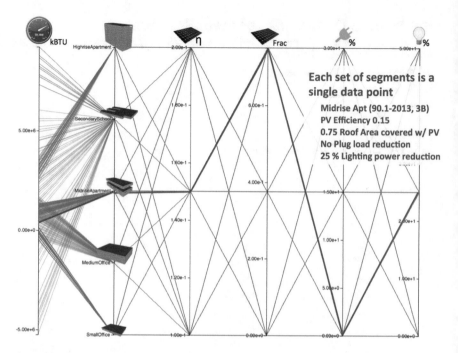

**Fig. 7.36** Net zero building study visualized as a parallel coordinate plot

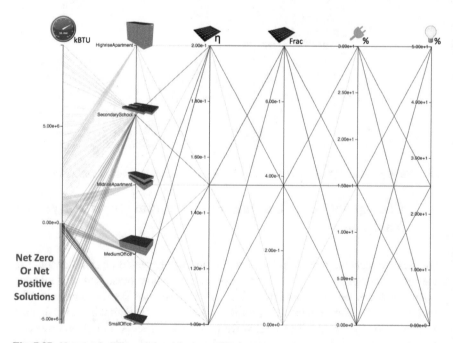

**Fig. 7.37** Net zero building study with output filtering

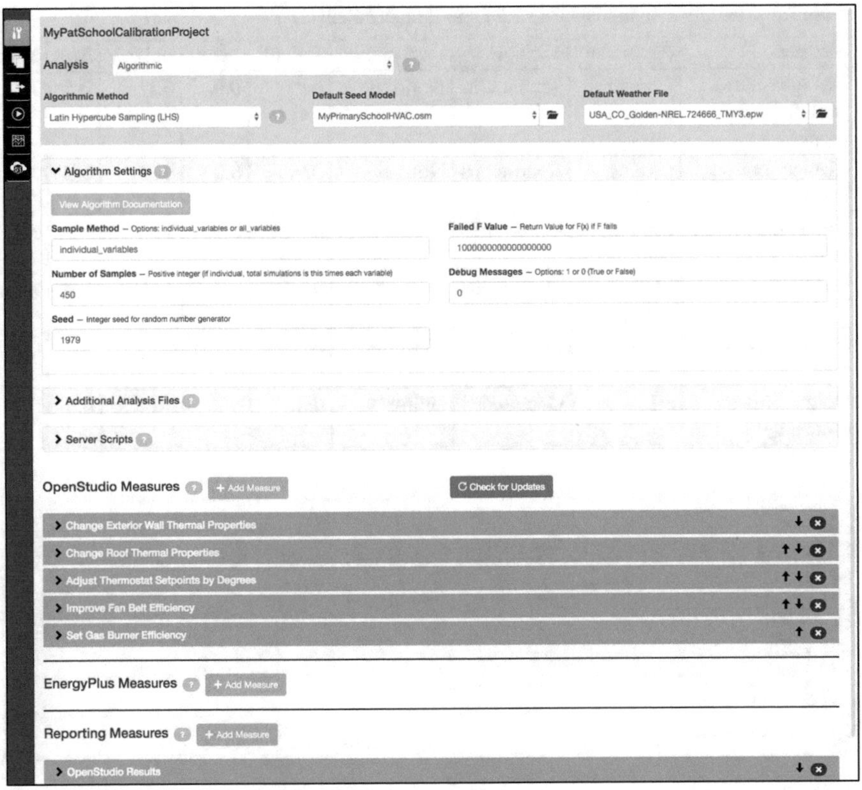

**Fig. 7.38** Setting up an LHS sampling problem with our school model

calibration. While we will focus on a narrow application of these algorithms, rest assured they may be used to solve a wide range of design problems.

We will continue to work with our Model from Chap. 4. Launch PAT and begin by:

1. Creating a new project called "MyPatSchoolCalibrationProject."
2. On the Analysis (⬛) Tab, set the Analysis mode to Algorithmic.
3. Set the Algorithmic Method to "Latin Hypercube Sampling (LHS)."
4. Add *MyPrimarySchoolHVAC.osm* as your default SEED Model.
5. Add USA_CO_Golden-NREL_724666_TMY3.epw as your default weather file.
6. Under algorithm settings, set the number of Samples to 75.[17]
7. Add the OpenStudio and Reporting Measures shown in Fig. 7.38.
8. Specify the Measure arguments as variables as shown in Table 7.1.
9. Proceed to the Outputs (⬛) Tab and add Outputs as shown in Fig. 7.39.
10. Save your project before proceeding to the Run (⬛) Tab.

---

[17]You may choose to use fewer samples for faster simulations or if you have fewer workers available. Our analysis will produce 6 × 75 or 450 data points.

**Table 7.1**  Variable assignments for LHS school model analysis

| Measure | Argument | Variable | Distribution | Default | Mean | Std dev | Min | Max |
|---|---|---|---|---|---|---|---|---|
| Change exterior wall thermal properties | r_value_ mult | Continuous | Normal | 0.9 | 0.9 | 0.1 | 0.5 | 1.5 |
| Change roof thermal properties | r_value_ mult | Continuous | Normal | 0.9 | 0.9 | 0.05 | 0.5 | 1.5 |
| Adjust thermostat setpoints by degrees | cooling_ adjustment | Continuous | Triangle | 0 | 0 | 1 | −3 | 3 |
| | heating_ adjustment | Continuous | Triangle | 0 | 0 | 1 | −3 | 3 |
| Improve fan belt efficiency | fan_eff | Continuous | Uniform | 0 | – | 5 | −40 | 0 |
| Set gas burner efficiency | eff | Continuous | Uniform | 0.8 | – | 0.1 | 0.6 | 0.9 |

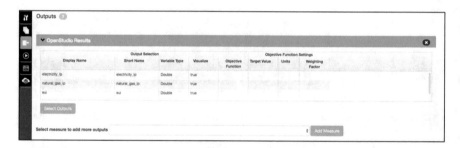

**Fig. 7.39**  Setting up outputs for LHS sampling problem with our school model

On the Run (◼) Tab select "Run on Cloud" and specify the Remote Server Type you will be using. Most users will select Amazon Cloud and set up their AWS run as discussed in Sect. 7.5.2. Configure your settings and connect to the server. Once connected, press the ▢Run Entire Workflow▢ Button to start the LHS analysis. The screenshot in Fig. 7.40 shows an analysis in process on a dedicated remote server.

> **Tip:** Before Running a large analysis, the authors always recommend testing a small version of it. You can do this by running a small number of LHS samples (1 or 2), or temporarily switching to the "Pre Flight" or "Single Run" algorithms.

Once the analysis is running, click the ▢View Server▢ Button to open the OpenStudio Server console in a web browser. The server's web page should look like Fig. 7.41 while the LHS analysis is in progress.

Clicking the ▢View Analysis▢ Button opens this particular LHS analysis' page shown in Fig. 7.42. This particular screenshot shows the page after all 450 data points have

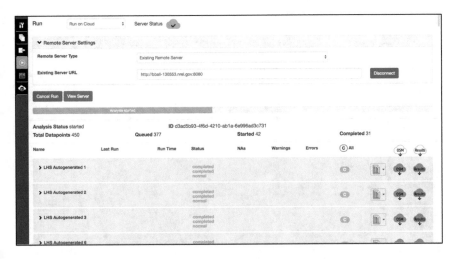

**Fig. 7.40** LHS sampling analysis in progress for school model problem

## OpenStudio Cloud Management Console

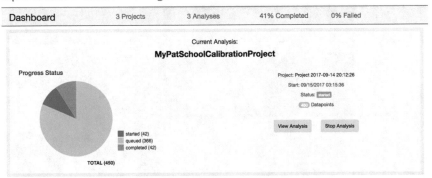

**Fig. 7.41** OpenStudio Server reporting progress of LHS analysis for school model problem

been simulated, however interim results and plots may be available while the analysis is in progress. One such result is the variable summary shown in Fig. 7.43. This page includes plots allowing us to verify the variable distributions we requested.

Once the LHS analysis is complete, take some time to explore the visualizations and data point information available via the links shown in Fig. 7.42. In addition to the individual data points, which include the Model files, EnergyPlus results, HTML reports, and more; the CSV and R Data Frame Results are useful downloads from the main analysis page that allow you to further explore the relationships between these inputs and outputs.

One particularly useful feature of OpenStudio Server is illustrated in Fig. 7.44, the parallel coordinate plot. As discussed in Sect. 7.6, this interactive tool allows the user to apply filter ranges to both the inputs and outputs to quickly identify a

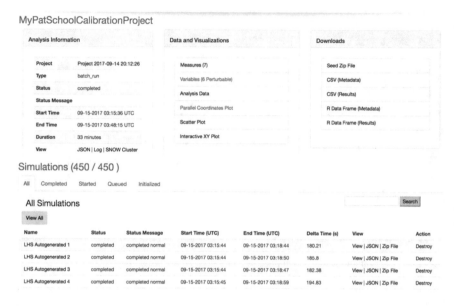

**Fig. 7.42** LHS analysis report page for school model problem

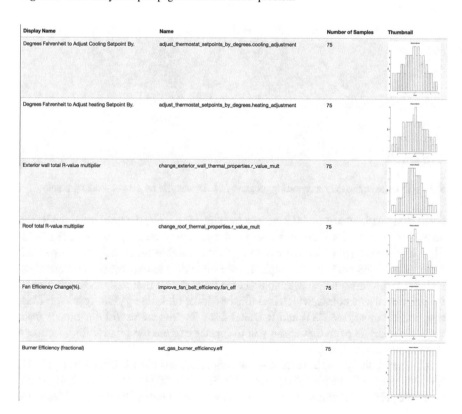

**Fig. 7.43** Variable distributions for LHS sampling of school model problem

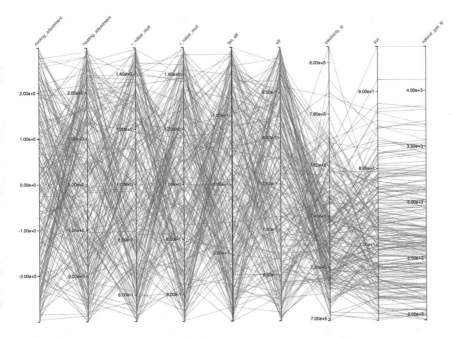

**Fig. 7.44** Parallel coordinate plot for LHS sampling of school model problem

particular data point out of the 450 we simulated that may be of interest. Without tools like this, finding solutions of interest can be a bit like looking for a needle in a haystack.

The LHS analysis we just completed contained a number of Measures related to the school's insulation, temperature setpoints, fan and gas burner efficiencies. The variable ranges we selected reflect a range of uncertain values that affect the energy performance of the building. For example, R value multipliers less than one reduce the effectiveness of the school's insulation, driving up energy consumption. Lower fan and burner efficiencies also degrade the performance of the building. The study produced a population of 450 schools reflecting variation in six dimensions. Obviously, this can be extended to encompass more characteristics related to school construction, Space loads, occupancy schedules, and HVAC systems. As more degrees of freedom (Measure variables) are added to such an analysis, the variation in energy performance will continue to increase.

Previous chapters have treated our primary school as a new building with Measures used to model design features that improve performance. However, we can also use Measures to approximate degraded performance due to host of circumstances like aging, poor construction, commissioning, or operation. This leads us to the next exercise where we will consider the hypothetical case that our school is not a new building but intended to model an existing school. The fundamental question we will attempt to answer: "If our school is best described by one of the 450 we simulated above, which one is it most likely to be?"

## 7.8   Checkpoint Eleven: Calibration via Sensitivity Analysis and Optimization

Modeling existing buildings presents different challenges than analyzing new building designs. When modeling for new construction, the design team and analyst may have significant leeway in changing the form, fabric, and sometimes even the function of a building, its Spaces, and systems. Aside from general performance goals, there are likely few expectations for the exact energy performance of a building, and comparative assessment of EE Measures is sufficient to drive design decisions; not so with an existing building. When faced with a decision to install a Measure, the owner of an existing building owner has existing utility bills as a benchmark of performance, and a natural expectation that EE investment will reduce those bills sufficiently to pay for themselves over some period of time. This raises the bar for the accuracy of modeled predictions and brings us to the topic of Model calibration.

Model calibration is the process of adjusting a Model's inputs to align the outputs with measured data. While we could consider measuring many aspects of building performance and "tuning" a Model to best represent them, we are primarily concerned with energy consumption. Energy consumption data may be provided in the form of monthly utility billing data, interval data from so-called "smart meters," or even finely grained time-series information recorded via data logger. OpenStudio has a means of adding interval consumption data to a Model for the purpose of calibration.

Pretend that our school was actually built in 1992 in Golden, CO. We have monthly electricity and gas consumption billing data from our local utility for the year 2017. Further, we have an EPW that represents the Actual Meteorological Year (AMY) weather for that year.[18] Open your trusty MyPrimarySchoolHVAC Model in the OpenStudio Application once more. On the Site (⬛) Tab make sure you are using the same weather file you used in the previous exercise. In the section labeled "Select Year by," check the Calendar Year box and set the year to 2017.

Next click on the [Utility Bills] Sub-Tab and use the ⬛ Button to add an Electric Utility Bill to your Model. Name the Bill "Electric," and add new billing periods as needed. Enter the monthly consumption data as shown in Figs. 7.45 and 7.46.

Add a second bill named "Gas" with the monthly consumption data shown in Fig. 7.47. Take care to enter the gas consumption data using the correct units of MBtu. Save your Model when you are finished and exit the OpenStudio Application.

So now we've added the "measured" data that we will judge how well calibrated our Model is. The next question we must ask ourselves, what variables do we need to modify to perform the calibration? As we have seen, a complete energy Model contains many thousand input parameters. Where do we even start knowing which ones to modify to align modeled and measured performance?

---

[18] It is important to note that model calibration should <u>always</u> be performed with AMY data corresponding to the period that the available calibration data was obtained. Using TMY data to calibrate a model is like mixing apples and avocados. That said, our "utility billing data" was generated from TMY weather conditions, so it is appropriate to use that weather file in this exercise.

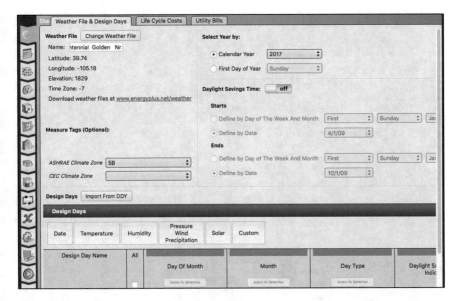

**Fig. 7.45**  Preparing to add electricity and gas consumption data to school model

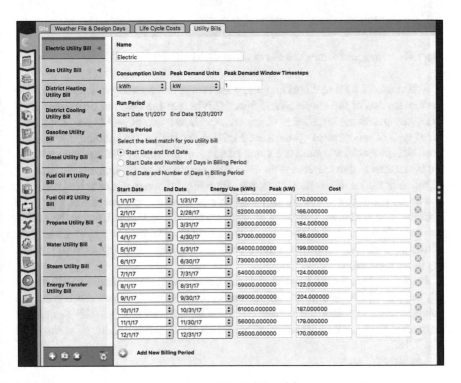

**Fig. 7.46**  Adding electricity consumption data to school model

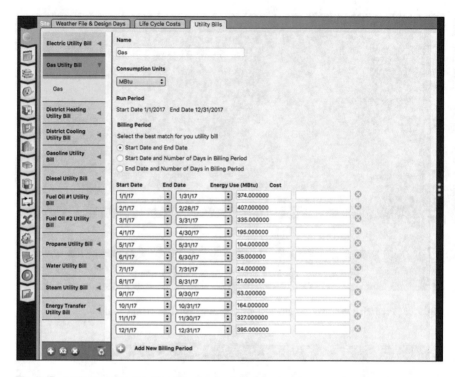

**Fig. 7.47** Adding gas consumption data to school model

When modeling a real building, there are things that are easy to know. What are the dimensions of the envelope and Spaces? What are the hours of operation? How many computers are installed in classroom, and how much power do they consume? What type of mechanical systems are installed for heating and cooling? Whenever possible, the modeler should use known data to inform inputs. Calibration variables are best suited to parameters the modeler is <u>uncertain</u> about. Equipment efficiencies, effective R values, infiltration rates, etc. are rarely known quantities. Nevertheless, one could claim that there are still hundreds of uncertain parameters in any given building. The next algorithm we will examine, the Morris Method, is very helpful in determining which variables most affect aspects of a building's performance. This allows us to identify and remove those variables that have little impact, while retaining those that matter.

In PAT, replace the old Seed Model with the one we just modified. You can manually copy the OSM into your PAT project's seed sub-folder or you can use the file menu in PAT to select the revised Model. On the Analysis (▣) Tab switch the Algorithm from LHS to Morris Method and set the Algorithm Settings as shown in Fig. 7.48.

Launch PAT and:

1. Load the "MyPatSchoolCalibrationProject."
2. Set the Algorithmic Method to "Morris Method."

3. Add your **newly modified** *MyPrimarySchoolHVAC.osm* as your default SEED Model.[19]
4. Under algorithm settings, set r to 10 and Levels to 30 as shown in Fig. 7.48.
5. Add the "Calibration Reports Enhanced" Reporting Measures shown in Fig. 7.48.
6. Proceed to the Outputs (▣) Tab and add Outputs as shown in Fig. 7.49.[20]
7. Save your project before proceeding to the Run (▣) Tab.

The most significant changes in the Project include the choice of algorithm and the addition of "performance objectives" that are computed by the Calibration Reports Measure using the consumption data we added to the Model. Among other things, the Measure produces two primary quantities that are indicative of the "goodness of fit" for a given simulation run. They are the net mean bias error (NMBE) and coefficient of variation of the root mean squared error (CVRMSE). These are defined according to:

$$NMBE = \frac{100}{\bar{y}} \times \frac{\Sigma\left(y_i - \hat{y}_i\right)}{n} \quad CVRMSE = \frac{100}{\bar{y}} \times \sqrt{\frac{\Sigma\left(y_i - \hat{y}_i\right)^2}{n}}$$

where:

$y_i$ is the $i$th measured data point,
$\hat{y}_i$ is the $i$th simulated data point,
$\bar{y}$ is the mean of the measured data points, and
$n$ is the total number of data points.

For our analysis, the Measure will produce NMBE and CVRMSE values based on the errors in monthly electricity and gas consumption resulting in four different performance objectives. Ideally, these will be zero for perfectly matched analytical and empirical values. In practice, these values can be minimized, but will never equal zero due to measurement and modeling errors. ASHRAE has published Guideline 14, which prescribes tolerances for NMBE and CVRMSE for building calibration.[21] A "good" calibration will have a NMBE less than 10% and CVRMSE less than 30%. The Calibration Reporting Measure also checks against these threshold values and reports pass/fail results. We will make use of these metrics for both the Morris Method and subsequent optimization of variables that minimize them.

Connect to the OpenStudio Server and run the Morris Method analysis. As before, you can open the Server window in a web browser to monitor the process and interact with interim results. The completed analysis should look like Fig. 7.50. The Morris Method is an algorithm that offers an additional download on the Project's page called *Algorithm Results.zip*. Download this file and examine its contents.

---

[19]You can manually copy the OSM into your PAT project's seed sub-folder or you can use the file menu in PAT to select the revised model, but **do not forget this step**.

[20]You may choose to leave the OpenStudio Results outputs in your project. We have removed them to minimize server reporting clutter.

[21]ASHRAE Guideline 14–2014: Measurement of Energy, Demand and Water Savings, ASHRAE, 2014.

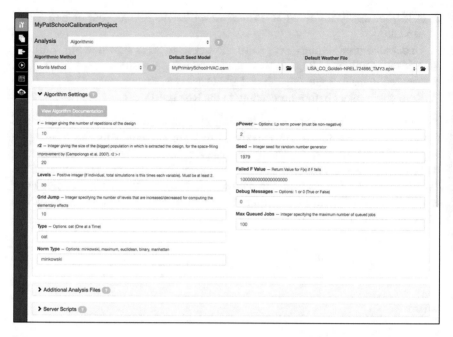

**Fig. 7.48** Setting up a Morris Method problem with our school model

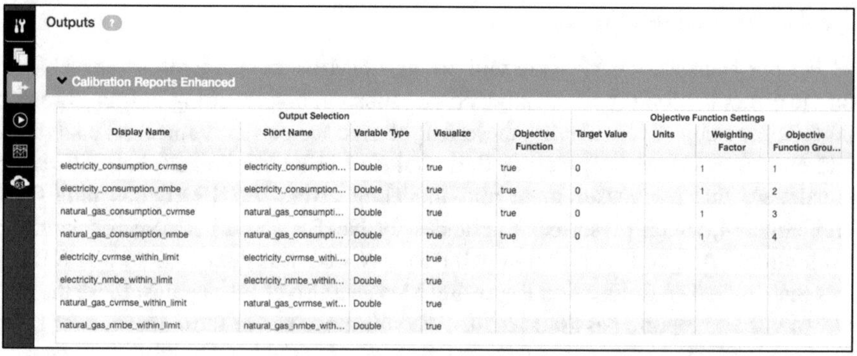

**Fig. 7.49** Setting up outputs for Morris Method problem with our school model

*Algorithm Results.zip* contains a number of pre-generated plots based on your choice of variables and the NMBE and CVRMSE objective functions. Figures 7.51 and 7.52 contains two sets of four plots related to gas and electric NMBE and CVRMSE. Figure 7.51 shows the "elementary effects" of each input variable on the output of interest. $\mu^*$ shown in Fig. 7.52 is the mean value of the absolute value of the elementary effects and represents the relative sensitivity of an output of interest to all input variables. Variables with comparatively large $\mu^*$ values are very significant in shaping an output, whereas $\mu^*$ equal to zero has no relationship.

# OpenStudio Cloud Management Console

MyPatSchoolCalibrationProject

| Analysis Information | | Data and Visualizations | Downloads |
|---|---|---|---|
| Project | Project 2017-09-14 21:37:31 | Measures (7) | Seed Zip File |
| Type | morris | Variables (6 Perturbable) | CSV (Metadata) |
| Status | completed | Analysis Data | CSV (Results) |
| Status Message | | Parallel Coordinates Plot | R Data Frame (Metadata) |
| Start Time | 09-15-2017 04:40:43 UTC | Scatter Plot | R Data Frame (Results) |
| End Time | 09-15-2017 04:52:33 UTC | Interactive XY Plot | Algorithm Results Zip |
| Duration | 12 minutes | | |
| View | JSON \| Log \| SNOW Cluster | | |

## Simulations (140 / 140 )

All   Completed   Started   Queued   Initialized

### All Simulations

Search

Toggle Obj Functions

View All

| Name | Status | Status Message | Start Time (UTC) | End Time (UTC) | Delta Time (s) | View | Action |
|---|---|---|---|---|---|---|---|
| API Generated 2017-09-15 04:41:12 +0000 | completed | completed normal | 09-15-2017 04:41:13 | 09-15-2017 04:44:33 | 200.01 | View \| JSON \| Zip File \| Bar Chart \| Radar Plot | Destroy |
| API Generated 2017-09-15 04:41:13 +0000 | completed | completed normal | 09-15-2017 04:41:13 | 09-15-2017 04:44:25 | 192.41 | View \| JSON \| Zip File \| Bar Chart \| Radar Plot | Destroy |
| API Generated 2017-09-15 04:41:13 +0000 | completed | completed normal | 09-15-2017 04:41:13 | 09-15-2017 04:44:27 | 194.37 | View \| JSON \| Zip File \| Bar Chart \| Radar Plot | Destroy |
| API Generated 2017-09-15 04:41:13 +0000 | completed | completed normal | 09-15-2017 04:41:13 | 09-15-2017 04:44:23 | 190.08 | View \| JSON \| Zip File \| Bar Chart \| Radar Plot | Destroy |
| API Generated 2017-09-15 04:41:13 +0000 | completed | completed normal | 09-15-2017 04:41:13 | 09-15-2017 04:44:22 | 188.76 | View \| JSON \| Zip File \| Bar Chart \| Radar Plot | Destroy |
| API Generated 2017-09-15 04:41:13 +0000 | completed | completed normal | 09-15-2017 04:41:13 | 09-15-2017 04:44:28 | 195.02 | View \| JSON \| Zip File \| Bar Chart \| Radar Plot | Destroy |
| API Generated 2017-09-15 04:41:13 +0000 | completed | completed normal | 09-15-2017 04:41:13 | 09-15-2017 04:44:29 | 195.67 | View \| JSON \| Zip File \| Bar Chart \| Radar Plot | Destroy |
| API Generated 2017-09-15 04:41:13 +0000 | completed | completed normal | 09-15-2017 04:41:13 | 09-15-2017 04:44:26 | 192.68 | View \| JSON \| Zip File \| Bar Chart \| Radar Plot | Destroy |
| API Generated 2017-09-15 04:41:13 | completed | completed | 09-15-2017 | 09-15-2017 | 188.68 | View \| JSON \| Zip File \| Bar Chart \| Radar Plot | Destroy |

**Fig. 7.50** Completed Morris Method analysis for school model problem

A few observations for our school Model are apparent. First, the most significant variables impacting gas consumption are not necessarily the same as those effecting electricity use. For example, the cooling setpoint change was the biggest driver of electricity consumption variability. For gas consumption, the heating setpoint was the most significant variable. Does this make sense?

Second, the burner efficiency has no impact on either gas or electricity performance metrics. This might seem surprising but open up the seed Model and examine its HVAC system to see if this outcome is reasonable. Assuming you have satisfied yourself that it has no impact, this variable (and Measure) may be removed from our Project.

What about the other variables? If we were only concerned about gas consumption, we might consider removing cooling setpoint as a source of uncertainty in our calibration. However, this would have a serious impact on our ability to shape the electricity consumption of our Model. The converse is true for the heating setpoint, so we will leave both of these in our project. Again, looking at gas consumption, one

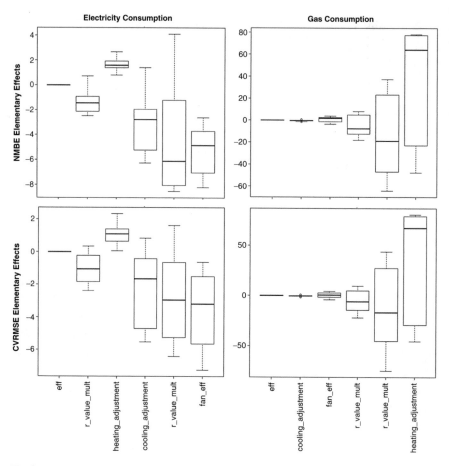

**Fig. 7.51** Morris Method elementary effects box plots for school model problem

might consider removing fan efficiency as a variable. Once again though, this would have a detrimental effect on our ability to shape electric consumption, as it is the second most significant elementary effect. Based on these considerations, we are probably wise to only prune the burner efficiency Measure from our calibration.

We are now going to shift from performing a sensitivity analysis of our four performance metrics to allowing SPEA2 to utilize the remaining Five variables to minimize NMBE and CVRMSE for gas and electricity consumption. Back in PAT:

1. Delete the burner efficiency Measure from the Project.
2. Set the Algorithmic Method to "SPEA2."
3. Under algorithm settings, set Number of Samples to $40^{22}$ and Generations to 2 as shown in Fig. 7.53.
4. Save your project before proceeding to the Run (▣) Tab.

---

[22] Algorithms like SPEA2, PSO, Rgenoud, etc. will make the most cost-effective use of your server and workers by setting the number of samples to be equal to the number of worker nodes available. This minimizes the number of workers that sit idle between optimizer iterations.

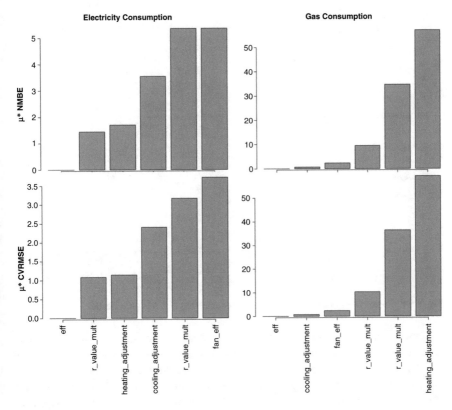

**Fig. 7.52** Morris Method μ* plots for school model problem

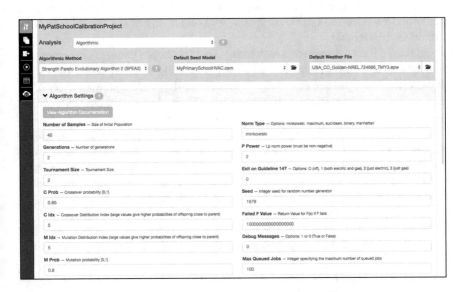

**Fig. 7.53** Setting up an SPEA2 optimization problem with our school model

MyPatSchoolCalibrationProject

| Analysis Information | | Data and Visualizations | Downloads |
|---|---|---|---|
| Project | Project 2017-09-14 15:26:08 | Measures (6) | Seed Zip File |
| Type | spea_nrel | Variables (5 Perturbable) | CSV (Metadata) |
| Status | completed | Analysis Data | CSV (Results) |
| Status Message | | Parallel Coordinates Plot | R Data Frame (Metadata) |
| Start Time | 09-14-2017 22:29:18 UTC | Scatter Plot | R Data Frame (Results) |
| End Time | 09-14-2017 22:39:19 UTC | Interactive XY Plot | |
| Duration | 10 minutes | | |
| View | JSON | Log | SNOW Cluster | | |

Simulations (120 / 120 )

All    Completed    Started    Queued    Initialized

All Simulations                                                                                    Search

Toggle Obj Functions

View All

| Name | Status | Status Message | Obj Func electricity_consumption_cvrmse | Obj Func electricity_consumption_nmbe | Obj Func natural_gas_consumption_cvrmse | Obj Func natural_gas_consumption_nmbe | Start Time (UTC) | End Time (UTC) | Delta Time (s) | View | Action |
|---|---|---|---|---|---|---|---|---|---|---|---|
| API Generated 2017-09-14 22:29:45 +0000 | completed | completed normal | 2.6657159812789994 | 0.8273049204717144 | 5.607111901414743 | 4.263128672867241 | 09-14-2017 22:29:45 | 09-14-2017 22:32:40 | 175.46 | View | JSON | Zip File | Bar Chart | Destroy |

**Fig. 7.54**  School model calibration project with completed SPEA2 optimization

Run the analysis and jump over to the OpenStudio Server console in your web browser (Fig. 7.54). This project is smaller and faster than the previous two by virtue of having removed a variable and the fact that the algorithm is focused on achieving a specific objective rather than studying the school Model's performance space more generally. Note in this screenshot we have clicked on the ▭ Button just above the data point list to display the four objective functions. Scrolling through the points reveals a range of good (and bad) solutions that the optimizer discovered on its way to the "best" combination of input variables.

Figure 7.55 uses the server's interactive XY plot feature to examine a trajectory of solutions that SPEA2 explored on its way to finding minimal values of NMBE and CVRMSE for gas and electricity.[23] Out of curiosity, you might want to try some of the other optimizers to see how they perform. Figure 7.56 shows how PSO arrives at a similar solution.

We have annotated these plots with regions showing data points that satisfy ASHRAE Guideline 14. This underscores two considerations. First, the solution space that we pulled our "real school" from in Checkpoint Nine was actually fairly small, and none of the variables we used to create this experiment were wildly divergent. Therefore, it's not that hard to find an answer that satisfies Guideline 14 for this simple problem.

More importantly though is the fact that a multiplicity of solutions exist that satisfy the criteria. This is very typical of building Model calibration problems and

---

[23] Because of the dimensionality of the space the optimizer is traversing, we can only visualize slices of the solution in plots like these.

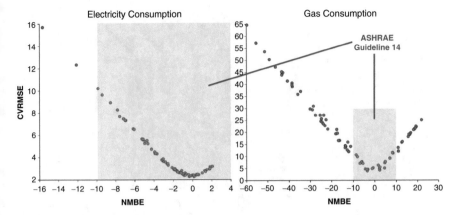

**Fig. 7.55** SPEA2 optimizer results for school model calibration problem

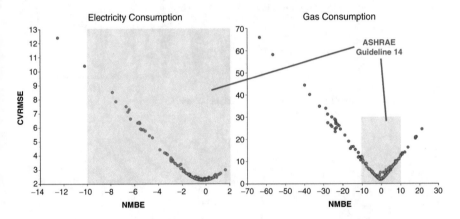

**Fig. 7.56** PSO optimizer results for school model calibration problem

becomes more pronounced when the number of uncertain variables increases while the number of dependent outputs (NMBE and CVRMSE) remains very small. These are known as mathematically underdetermined problems, and they result in multiple possible solutions yielding similar outcomes. While an unimpeachable, mathematical solution may elude us; we can bring some human expertise into play to narrow the range of potential solutions.

OpenStudio Server's interactive parallel coordinate plot can be helpful in exploring solution spaces and identifying data points of interest (Fig. 7.57). You may have guessed that the four "within_limit" outputs shown on the far right of this plot correspond to solutions that achieved ASHRAE Guideline 14 electricity or gas compliance for NMBE or CVRMSE. Use your mouse to filter where these outputs are equal to one and the corresponding metric falls within tolerances (Fig. 7.58). The remaining blue "threads" represent simulation data points that fall within both green boxes shown in Fig. 7.55.

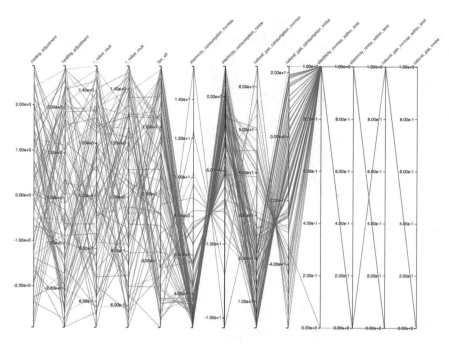

**Fig. 7.57**  Parallel coordinate plot of all SPEA2 data points

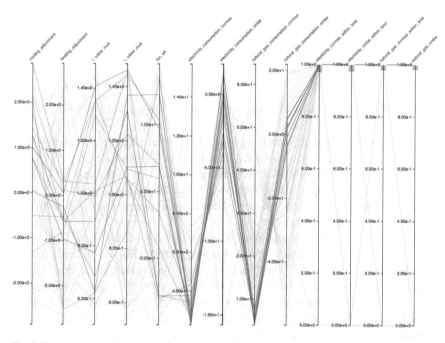

**Fig. 7.58**  Parallel coordinate plot of SPEA2 results filtered for ASHRAE Guideline 14

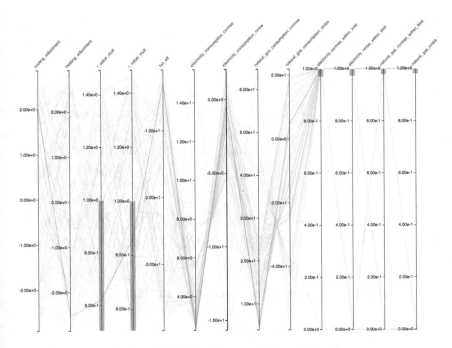

**Fig. 7.59** Additional filtering of SPEA2 results on parallel coordinate plot

That helped a bit, but there are still a large number of potential solutions. To refine our search further, we need to make use of one seemingly insignificant detail that you may have overlooked at the beginning of this exercise – the school was built in 1992.

How can the vintage of the building help us narrow our search? A seasoned energy modeler would recall one of the very first assumptions that we made back in Chap. 2 when we first built our school Model. In that exercise, we assigned a Construction Set corresponding to ASHRAE 90.1-2010. This represents a more stringent level of the energy code than when our hypothetical school was built nearly two decades earlier. It's quite likely that our school has less insulation than how we modeled it. In our study, that would correspond to data points with R value multipliers less than one. Applying an appropriate filter to the wall and roof "r_value_mult" variables produces Fig. 7.59. Abracadabra!

Below the parallel coordinate plot is a list of all data points that meet the filter requirements. As we continue to filter results, the list has winnowed down to less than a handful of points. In fact, these points are identical solutions that the optimizer arrived at from different directions. Selecting one of them allows us to see the specific Measure variable values that define the data point. The data point page also allows convenient download of the specific Model, simulation results, and associated reports shown in Fig. 7.60.

Output from the Calibration Report for this data point is one of the available report links and allows us to quickly visualize the performance of this particular Model relative to our school's "measured" consumption data. Figure 7.61 compares one such report for a bad solution with the SPEA2 solution that met our criteria.

**Fig. 7.60** An SPEA2 optimized solution for the school calibration problem

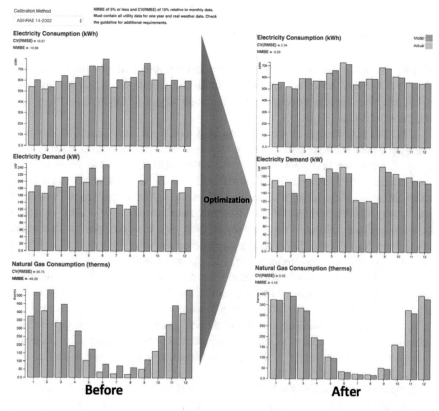

**Fig. 7.61** School model calibration reports pre and post SPEA2 optimization

This exercise attempted to provide some experience and insight in using PAT and OpenStudio Server to solve a common problem that faces building modelers focusing on retrofit of existing buildings. Given the huge number of uncertain variables associated with any building, these are fundamentally difficult problems to solve. It is important to approach the tools with realistic expectations, and to leverage common sense, experience, and even intuition when trying to arrive at the "best" Model for a particular building.

## 7.9   Additional Exercises

There are a number of potential exercises involving the Model you created for "Additional Exercises" in Chap. 4. A few to consider include:

1) Learn more about uncertainty in predicted Model performance.

  - Consider sources of uncertainty in your model and select Measures for the BCL that allow you to modify related inputs: e.g. electric load power, infiltration, HVAC system performance, etc.
  - Create continuous or discrete variables for key parameters in these Measures and use them in an LHS problem to see how variation impacts annual energy use and peak power consumption.
  - Download R or Excel data from OpenStudio Server to plot EUI distributions for the Models in your sample population.

2) Identify some of the key parameters your Model is most sensitive to.

  - Expand upon the previous analysis by using Morris, Sobol, or FAST99 to identify parameters that significantly impact annual energy use, peak power consumption, or other outputs of potential interest.

3) Calibrate your Model using available consumption data.

  - Obtain electric or gas consumption data from the building owner or facility manager.
  - Obtain an AMY weather file for your region.
  - Calibrate your Model using the process described in Checkpoint Eleven.

4) Optimize efficiency Measures for your calibrated Model.

  - Set up an objective function[24] to assess the relative goodness of a collection of efficiency measures.
  - Select Measures and input ranges for your optimization problem.
  - Experiment with one or more optimization algorithms and compare the solutions they produce.

---

[24] Give some thought to definition of an objective function. For example, optimizing on annual energy use alone will result in a building with no windows, maximum insulation, ultra-high efficiency HVAC systems, etc. Add additional objectives (e.g. cost, comfort, or competing factors like heating and cooling energy) to create more realistic optimization problems.

# References

Barricelli N (1957) Symbiogenetic evolution processes realized by artificial methods. Methodos 9:143–182

Belding T (1995) The distributed genetic algorithm revisited. In: Eshelman LJ (ed) Proceedings of the sixth international conference on genetic algorithms. Morgan Kaufmann, San Francisco, CA, pp 114–121

Deb K, Pratap A, Agarwal S (2002) A fast and elitist multi-objective genetic algorithm: NSGA-II. IEEE Trans Evol Comput 6:182–197

Ephramac (2017) 2D-Partikelschwarm sucht globales Minimum, 12 Jan 2017. https://commons.wikimedia.org/wiki/File:ParticleSwarmArrowsAnimation.gif

Fraser A (1957) Simulation of genetic systems by automatic digital computers. Aust J Biol Sci 10:484–491

https://www.r-project.org/

Kennedy J, Eberhart R (1995) Particle swarm optimization. Proc IEEE Int Conf Neural Netw IV:1942–1948

King R, Deb K, Rughooputh H (2010) Comparison of NSGA-II and SPEA2 on the multi-objective environmental/economic dispatch problem. Univ Mauritius Res J 16:485–511

McKay M, Beckman R, Conover W (1979) A comparison of three methods for selecting values of input variables in the analysis of output from a computer code. Technometrics 21:239–245

Mebane W Jr, Sekhon J (2011) Genetic optimization using derivatives: the RGenOUD package for R. J Stat Softw 42(11):1–26. https://www.jstatsoft.org/article/view/v042i11

Morris M (1991) Factorial sampling plans for preliminary computational experiments. Technometrics 33:161–174

Nelder J, Mead R (1965) A simplex algorithm for function minimization. Comput J 7:308–313

Saltelli A, Tarantola S, Chan K (1999) A quantitative, model independent method for global sensitivity analysis of model output. Technometrics 41:39–56

Sobol I (2001) Global sensitivity indices for nonlinear mathematical models and their Monte Carlo estimates. Math Comput Simul 55:271–280

Zitzler E, Thiele L (1998) An evolutionary algorithm for multi-objective optimization: the strength Pareto approach. Technical report 43, Computer Engineering and Networks Laboratory (TIK), Swiss Federal Institute of Technology (ETH) Zurich

# Chapter 8
# Daylighting

## 8.1 Introduction

After mastering the basics of energy modeling with OpenStudio and EnergyPlus, there are still many advanced modeling features left to explore. A few include:

- Radiant heating and cooling systems that can achieve deeper savings for some climates and building types,
- Distributed energy resources and storage systems may be added to design net zero or off grid buildings,[1]
- Large refrigeration for grocery store and warehouse models,
- Airflow network models may be added for improved infiltration and natural ventilation assessment, and
- EnergyPlus' EMS scripting language which allows users to model custom building controls and sequences of operation.

These features and many more, integrated within EnergyPlus are beyond the scope of this introductory text. However, the approaches we have discussed for comparative analysis of building design features is equally applicable when considering these advanced topics. The EnergyPlus Input Output Reference is the definitive reference for learning more about EnergyPlus' full range of capabilities.

In addition, OpenStudio can facilitate interoperability and co-simulation with other engines and solvers that are designed to model specific building or grid-related phe-

---

[1] Distributed energy resources include solar hot water, photovoltaic, and wind turbine, and generator systems. A few examples of thermal and electric storage include phase change construction materials and mechanical systems or batteries.

The original version of this chapter was revised. A correction to this chapter can be found at https://doi.org/10.1007/978-3-319-77809-9_10

**Electronic Supplementary Material:** The online version of this chapter (https://doi.org/10.1007/978-3-319-77809-9_8) contains supplementary material, which is available to authorized users.

© Springer International Publishing AG, part of Springer Nature 2018
L. Brackney et al., *Building Energy Modeling with OpenStudio*,
https://doi.org/10.1007/978-3-319-77809-9_8

nomena in greater detail. One such program, and the subject of this Chapter, is Radiance.[2] While EnergyPlus can model daylighting and calculate associated energy savings, that isn't its primary function and the algorithms it uses to do so are simplified. Radiance has been purpose-built to model light propagation in buildings and is capable of calculating lighting metrics with far greater accuracy. OpenStudio pairs EnergyPlus with Radiance to deliver superior energy savings estimates for lighting-related ECMs.

## 8.2  Daylighting

Daylighting is a key design strategy used in low energy buildings, particularly offices, which tend to have high lighting loads and daytime occupancy. By making use of naturally available light entering the building through windows, skylights, and other apertures; electric lights may be dimmed or turned off when they are not needed. This strategy provides direct lighting energy savings. In addition, reducing lighting heat rejection can lower cooling loads for further HVAC energy savings. Beyond the energy benefits, many studies[3,4] have found that daylit spaces are more comfortable, pleasant, and healthy for occupants.

Creating a well daylit space requires careful design, far beyond the energy performance tradeoff of adding fenestration to provide daylight access while potentially reducing the envelope's thermal performance. Ensuring that uniform illumination is available for occupants throughout a space is a non-trivial exercise for daylit spaces. External shading or daylight redirecting devices are often required to prevent excessive glare that can distract or disturb occupants. Finally, the visual characteristics of a space include a certain *je ne sais quoi* that cannot be fully captured in the tabular output of an energy simulation.

As noted in Chap. 3, EnergyPlus does not perform detailed electric lighting illuminance calculations during simulations. Lighting loads and schedules are prescribed a priori, and the direct energy used by lighting is accounted for as well as the additional heat generated by lighting equipment. For this reason, EnergyPlus is not typically used to design and validate electric lighting layouts. However, EnergyPlus does have the ability to model the impact of daylighting controls, which may dim or turn off electric lights when sufficient natural light is available.

EnergyPlus uses a split flux method to calculate available daylighting as a combination of direct, externally reflected, and internally reflected light.[5] For points of interest inside the building, these components are related to the exterior solar insolation through daylight factors. Daylight factors are calculated for a variety of sun positions (e.g. once per month at each hour of the day) and then interpolated based

[2] Ward and Shakespeare (1998).

[3] https://cdn2.hubspot.net/hub/155785/file-18058478-pdf/docs/daylighting_research_-_us_government_report.pdf.

[4] http://www.sciencedirect.com/science/article/pii/S0378778806000624.

[5] http://bigladdersoftware.com/epx/docs/8-7/engineering-reference/time-step-daylighting-calculation.html.

on the actual sun position at any given simulation time step. This technique provides a first order approximation of the available daylight illuminance inside the building and is sufficient to estimate potential savings due to daylighting control strategies. The split flux method does not consider all exterior and interior reflections. Further, it cannot be used to provide detailed renderings of the scene. For these reasons, EnergyPlus is not often used to aid in detailed daylighting design.

A variety of other tools are available to assist in the design of electric and natural lighting systems. One such tool is the open source Radiance software package, which was developed with DOE support beginning in 1985. Radiance uses a detailed backward ray-tracing method to compute interior illuminance, capturing all interior and exterior reflections. This results in a more accurate simulation that may be used to predict electric and natural lighting levels for design purposes. Additionally, Radiance can be used to generate scene renderings useful in design. Like EnergyPlus, the Radiance package consists of command line tools with complicated input formats which are difficult for beginners to master.

The OpenStudio Object Model includes components necessary to perform daylighting simulation using EnergyPlus. Additionally, the Radiance Daylighting Measure has been developed to:

- Export an OpenStudio Model with daylighting elements for use by Radiance,
- Perform daylighting analysis using the Radiance engine,
- Post-process Radiance results to identify daylight availability within the Model, and
- Merge the availability data into the Model to inform EnergyPlus lighting schedules.

This gives a user the option to use either EnergyPlus or Radiance for daylighting simulations depending on their needs. OpenStudio Luminaire Objects are not presently exported to Radiance for electric lighting simulation, this is an anticipated future extension to the Measure.

### 8.2.1 Daylighting Controls

A Daylighting Control Object is required to enable daylighting calculations in OpenStudio. These Objects are placed in an OpenStudio Space. The specific location of the Object within a Space defines the point at which illuminance is calculated for the purpose of daylighting control. Daylighting Controls may be placed on a floor plan using the floorplan editor or in a detailed 3D space using the OpenStudio SketchUp Plug-in. Additionally, several OpenStudio Measures exist which automatically place Daylighting Control Objects in Spaces of interest.[6] An OpenStudio Model with Daylighting Control Objects may be simulated directly in EnergyPlus. These same Objects and the Model's geometry are also used when the Radiance Measure is included as part of a workflow, requiring no additional effort from the user.

---

[6]The AedgK12FenestrationAndDaylightingControls Measure (https://bcl.nrel.gov/node/39783) we used in Checkpoint Eight is one example.

While OpenStudio allows Daylighting Control Objects to point in arbitrary directions but EnergyPlus currently only allows these sensors to point up. Daylighting Control Objects do not currently support directionally sensitive inputs. Illuminance from all directions in front of the sensor is weighted equally. Given these limitations, current best practice is to place Daylighting Control Objects on the work plane instead of their actual location within the space, which is often on the ceiling. Used in this way, Daylighting Control Objects represent ideal daylight sensors.

In addition to position, Daylighting Control Object parameters describe the control strategy which will be used to dim or turn off lights in response to available daylight. As of OpenStudio 2.3.0, a given Daylighting Control Object cannot be used to control a specific light or group of lights. Instead, the user specifies up to two Daylighting Control Objects. Each Control Object controls a fraction of all Light and Luminaire Objects within a Thermal Zone. This restriction was originally put in place due to EnergyPlus limitations. Recently, EnergyPlus was upgraded to support less restrictive daylighting configurations, and future versions of OpenStudio may support these as well.

When used with the Radiance Measure, Daylighting Control Objects define the location at which ray tracing-based illuminance calculations will be performed. Based on the control strategy, the Measure modifies fractional lighting Schedules for Light and Luminaire Objects based on the daylight availability. Because these modified schedules already reflect the intended daylighting control action, the Measure then removes the Daylighting Control Objects prior to EnergyPlus simulation. This prevents EnergyPlus from taking additional action that would compromise the results of the Radiance analysis. Radiance time series analysis results are written to an SQL file in the same format as EnergyPlus results, allowing easy comparison of EnergyPlus and Radiance results with DView. Additional results from the Radiance simulation are available in the Measure's run directory.

## 8.2.2  Illuminance Maps

Daylighting Control Objects are used to measure daylight illuminance at a given point and to dim or turn off lights in response to available daylight. This is useful for estimating energy savings potential. However, it does not give a full view of the available daylight throughout the space. The Illuminance Map Object is useful for this purpose. Illuminance Map Objects specify a rectangular grid of points for which daylight illuminance is calculated at each time step using the same method as for Daylighting Control Objects. Illuminance Map Objects cannot be used to control Light and Luminaire Objects, the Illuminance Map is only used to provide output for visual inspection. EnergyPlus writes Illuminance Map output to a human readable CSV format. The deprecated OpenStudio ResultsViewer application (available in OpenStudio 2.2.0 and previous releases) allows the user to visualize Illuminance Map output in a graphical form. As of OpenStudio 2.3.0, the DView application, which has replaced OpenStudio ResultsViewer for output visualization, does not have the ability to visually display Illuminance Map output from EnergyPlus.

Similar to Daylighting Control Objects, the OpenStudio Radiance Measure operates on the same Illuminance Map Object that is used for EnergyPlus results. The OpenStudio Radiance Measure requests a daylight calculation for each point in the Illuminance Map. It then writes these results to a SQLite file in the same format that EnergyPlusdoes, allowing the user to visualize both results side by side using the OpenStudio ResultsViewer program. In addition to Illuminance Maps, the Radiance Measure can also be used to generate photo-realistic renderings of the entire scene.

### 8.2.3  Advanced Daylighting Objects

After mastering the Daylighting Control and Illuminance Map Objects, there are many other even more advanced Objects for daylight simulation. The Glare Control Object can be used to calculate glare at a specific location. This is useful to ensure that daylighting designs which provide sufficient daylight illuminance for tasks do not bring in too much daylight which causes distracting glare. The Shading Control Object is used to operate shading devices such as blinds, shades, or screens in response to daylight illuminance or glare. There are very detailed inputs for this Object (e.g. controlling the slat angle of blinds), the user is directed to the EnergyPlus Input Output Reference for details. Unlike Daylighting Control and Illuminance Map Objects, the OpenStudio Radiance Measure does not fully incorporate all of these advanced daylighting Objects. Users should take caution when using these Objects with the OpenStudio Radiance Measure.

## 8.3  Checkpoint Twelve: Daylighting Analysis

In this example, we will be performing a daylighting analysis on a portion of a Model imported from gbXML, an industry-supported schema for sharing building information between building energy design software tools.[7] Many software tools are capable of reading or writing gbXML files. OpenStudio is able to both read and write gbXML. The OpenStudio Application imports gbMXL files through the "File/ Import/gbXML" menu as shown in Fig. 8.1.

For this exercise, we will use a sample gbXML file available online. Open a web browser and navigate to https://github.com/GreenBuildingXML/Sample-gbXML-Files. Download the *gbXML_TRK.xml* example file by pressing "Clone or Download" and then select "Download ZIP." This gbXML file was exported from an Autodesk Revit MEP 2014 model and represents a mixed-use building with retail

---

[7] https://gbxml.org.

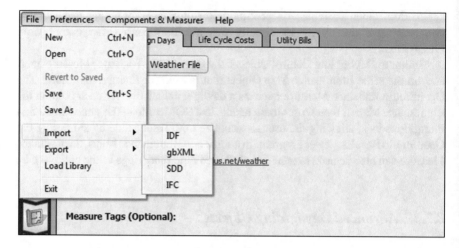

**Fig. 8.1** gbXML import within the OpenStudio Application

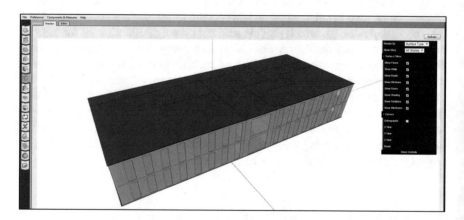

**Fig. 8.2** gbXML Model import previewed in the geometry Tab

on the first floor and a mix of office and gymnasium spaces on the second floor. A large two-story atrium serves as the main entrance and lobby. Once imported into the OpenStudio Application, the geometry can be viewed using the Geometry (▣) Tab as shown in Fig. 8.2.

After importing a gbXML file, non-geometric portions of the model such as space types, thermal zones, and HVAC may all be specified in the OpenStudio Application. However, the Application's floorplan editor cannot be used to edit potentially complex Models that have been imported from gbXML. For this reason, we introduce the OpenStudio SketchUp Plug-In for this exercise. Use of the SketchUp Plug-in requires that the user download and install SketchUp from https:// sketchup.com. SketchUp should be installed <u>before</u> installing OpenStudio, however

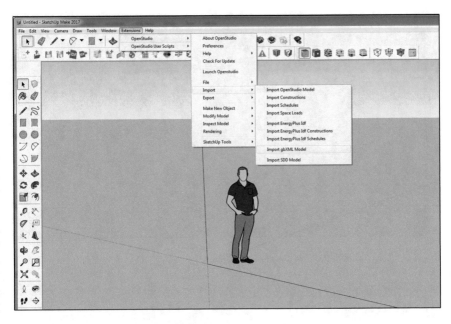

**Fig. 8.3** OpenStudio SketchUp Plug-in gbXML Import Menu

the OpenStudio installer can simply be run a second time if you did not previously install SketchUp. As of OpenStudio 2.3.0, SketchUp Make 2017 and SketchUp Professional 2017 are supported.

Once installed, the OpenStudio SketchUp Plug-in will automatically start when you launch SketchUp. The OpenStudio SketchUp Plug-in installs a variety of OpenStudio specific tools on the SketchUp toolbar. It will also create "Extensions/ OpenStudio" and "Extensions/OpenStudio User Scripts" menu items. We will only be using a small subset of the OpenStudio SketchUp Plug-in for this exercise. A more extensive reference guide to the Plug-In is available online.[8]

Start SketchUp and verify that the OpenStudio SketchUp Plug-in is loaded by checking for the "Extensions/OpenStudio" menu (Fig. 8.3). Navigate to "Extensions/ OpenStudio/Import/Import gbXML Model," select the *gbXML_TRK.xml* you downloaded in the file selection dialog.

After the gbXML file is loaded, you should see the Model shown in Fig. 8.4. Note that SketchUp's solid green axis shown in Fig. 8.4 is North and the solid red axis is East. Rotate around the Model to view all of the Spaces as shown in Fig. 8.5.

For this example, we are going to examine the daylighting in one Space only, the Space named "1 Atrium" which is associated with the Thermal Zone "Atrium". Press the ▪ Button, which renders the Model according to Thermal Zone as shown

---

[8] https://nrel.github.io/OpenStudio-user-documentation/reference/sketchup_plugin_interface.

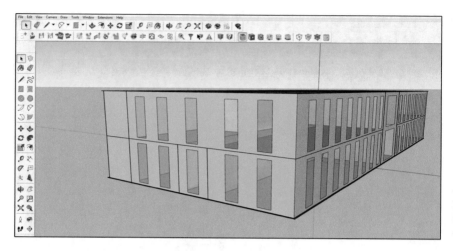

**Fig. 8.4** Initial Model import screen in SketchUp

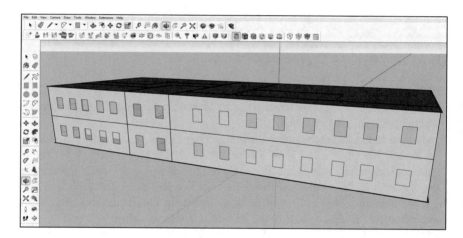

**Fig. 8.5** Rotated view of Model in SketchUp

Fig. 8.6. Use the Info Tool, selected using the ✎ Button, to mouse over each Thermal Zone to find the "Atrium" Zone. Click on each Space that is not part of the "Atrium" Thermal Zone and delete it with the Delete Key.

> Tip: The OpenStudio SketchUp Plug-in does <u>not</u> support SketchUp's undo (Ctrl-Z) feature. If you make a mistake while modifying your model, it is best to close SketchUp and start over. For this reason, "Save Early and Save Often" is a useful mantra when working with the SketchUp Plug-in.

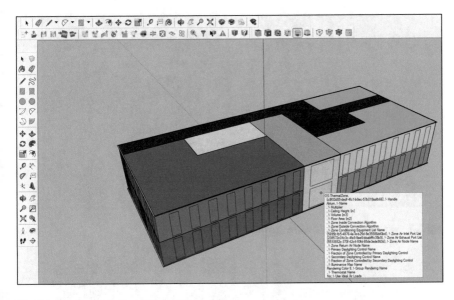

**Fig. 8.6** Rendering in SketchUp by Thermal Zone

**Fig. 8.7** Isolated Atrium
Space from Original Model

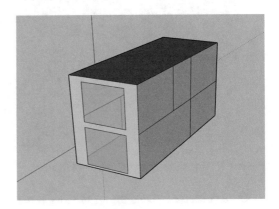

After successfully deleting all the other spaces in your Model, you will be left with the two story atrium as shown in Fig. 8.7. Heading our mantra, save the OpenStudio Model at this point.

> **Tip**: The OpenStudio SketchUp Plug-in's Save tool is <u>separate</u> from the native SketchUp Save command you access from the File menu or Ctrl-S. When working in the OpenStudio SketchUp Plug-in, use the OpenStudio ⊟ and ⬆ Buttons to save and load your Model. SketchUp's native SKP file format is not used by OpenStudio and can be ignored.

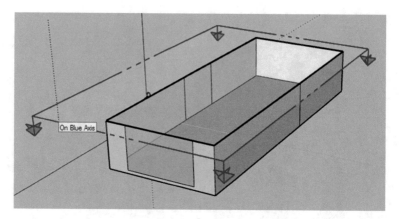

**Fig. 8.8** Section Cut of the Atrium Space

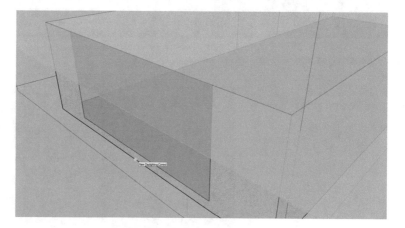

**Fig. 8.9** Placing a Daylighting Control in a Model

Once you have isolated the atrium Space and saved your OpenStudio Model, we will add a section cut to help us see "inside" the Space. The Section Cut Tool is part of the SketchUp Large Tool Set and is selected with the ⊕ Button. If you do not see it, select "View/Toolbars" and select "Large Tool Set". After clicking the ⊕ Button, click on the Space's roof to place the section cut. Use the Move tool selected with the ✤ Button to move the section cut up and down to see inside the space as shown in Fig. 8.8.

Once you have placed your section cut, switch back to the ✹ Tool and double-click on the atrium Space to open the related SketchUp Group for editing. All OpenStudio geometry for a given Space is created in a related SketchUp group, and geometry can only be modified when the SketchUp group is selected. Once the Space has been selected, click the ✺ Button to add a new Daylighting Control Object. Click on the floor in the middle of the space under the window to place the Daylighting Control as shown in Fig. 8.9.

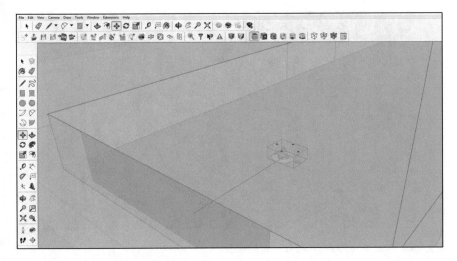

**Fig. 8.10** Moving a Daylighting Control Within a Model

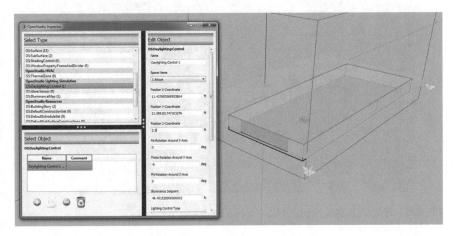

**Fig. 8.11** Inspecting Daylighting Controls

Once the Daylighting Control has been placed inside the Space, it can be moved with the ✦ Tool. The ✦ Tool is very useful in that it allows you to lock on a particular direction using the Shift Key. You can specify the distance to move the selected Object by entering a distance with units into the text box at the lower right-hand corner then pressing Enter. Move the Daylighting Control along the negative red axis (South) twelve feet as shown in Fig. 8.10.

Properties of the Daylighting Control Object may be inspected and edited using the ✜ Button as shown in Fig. 8.11. In addition to the position of the control sensor, the Illuminance Setpoint is another key parameter of the Daylighting Control Object. This is the illuminance level at which daylight illuminance is deemed sufficient to

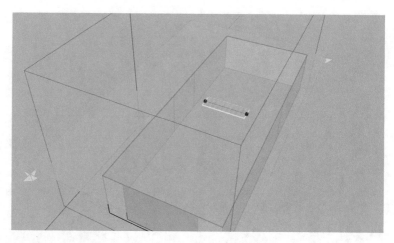

**Fig. 8.12**  Placing an Illuminance Map in a Model

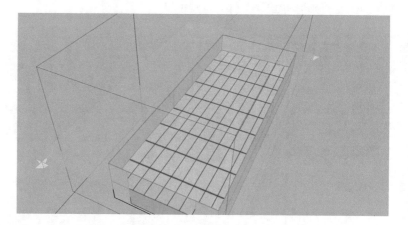

**Fig. 8.13**  Resizing an Illuminance Map within a Model

turn off electric lights. Because EnergyPlus is only concerned with the power and thermal characteristics of lighting, it is assumed that the lighting design is sufficient to provide this level of illumination without any available daylight.

After positioning our Daylighting Control Object, we also want to add an Illuminance Map to our Model using the ▦ Button. Illuminance Map Objects are placed in a similar way to Daylighting Control Objects. Once the Illuminance Map is placed it can be moved just like the Daylighting Control Object. The Illuminance Map Object is unique as the only OpenStudio Model Object that may be modified with the SketchUp Scale Tool. Select the Scale Tool with the ▦ Button and modify the Illuminance Map to cover the entire atrium as shown in Figs. 8.12 and 8.13.

Similar to the Daylighting Control Object, you can use the ⬚ Button to modify properties of the Illuminance Map Object as shown in Fig. 8.14.

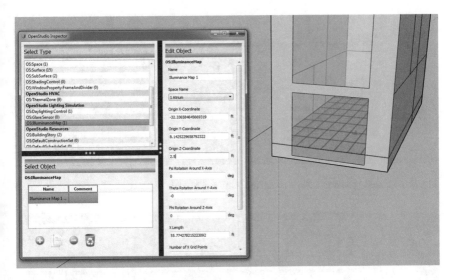

**Fig. 8.14** Inspecting the Illuminance Map

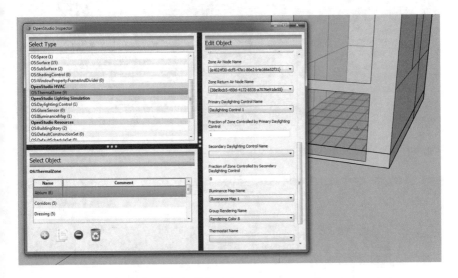

**Fig. 8.15** Inspecting the Atrium Thermal Zone

While the OpenStudio Inspector window is open, navigate to the Thermal Zone Object and select the atrium Thermal Zone. Check that the Daylighting Control and Illuminance Map we just placed are selected as the primary Daylighting Control and Illuminance Map for the atrium Thermal Zone (Fig. 8.15).

At this point, we have reduced our Model to only include the atrium Space and placed our Daylighting Control and Illuminance Map Objects. This is as far as we need to go in the OpenStudio SketchUp Plug-in. Save your OpenStudio Model

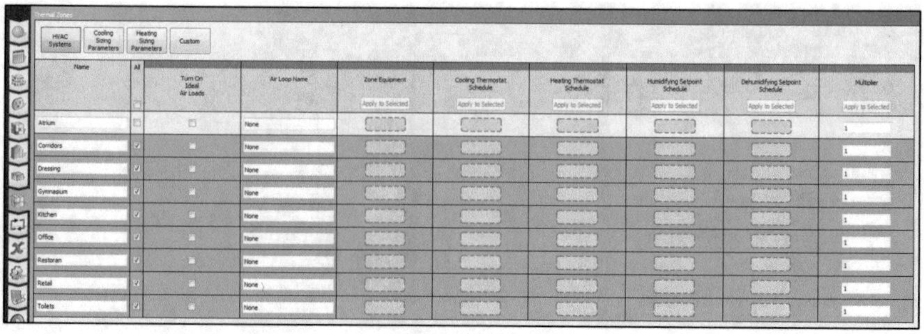

**Fig. 8.16**  Deleting Thermal Zones from the Imported Model

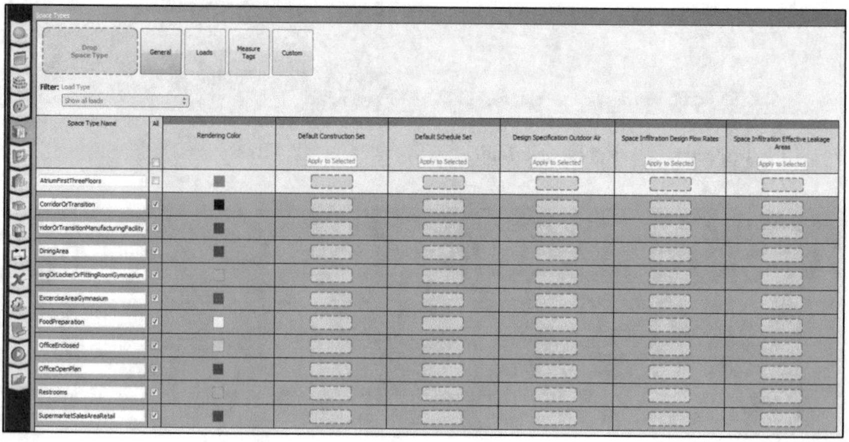

**Fig. 8.17**  Deleting Space Types from the Imported Model

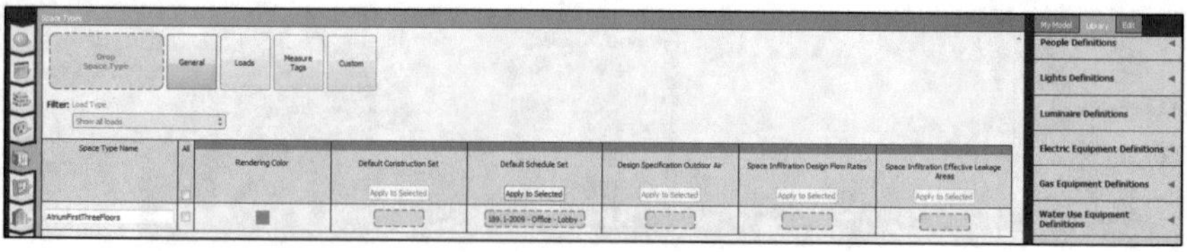

**Fig. 8.18**  Applying a New Schedule Set to the Imported Model

using the ▨ Button. **Do not use SketchUp's Save**. Switch to the OpenStudio Application and open the Model. The ♥ Button in the Plug-In allows you to do this with a single mouse click.

After opening your Model in the OpenStudio Application, set a weather file and import design days on the Site (▨) Tab. Then, navigate to the Thermal Zones (▨) Tab. Notice that deleting the non-atrium space geometry in the SketchUp Plug-in did not delete the associated Thermal Zones, which were imported from the original gbXML. Select all the Thermal Zones besides the atrium and delete them as shown in Fig. 8.16.

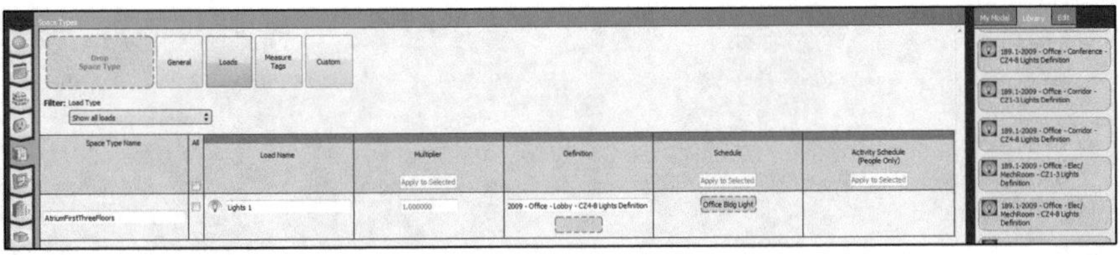

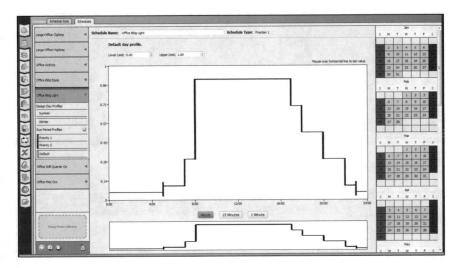

**Fig. 8.19**  Applying a Lighting Definition

**Fig. 8.20**  Reviewing the Lighting Schedules

Next, navigate to the Space Types (■) Tab. Notice that the other imported Space Types were not deleted either. Select and delete them as shown in Fig. 8.17.

The Space Types imported from this gbXML file did not include lighting loads or lighting schedules. Drag in the "189.1-2009 – Office – Lobby" Schedule Set and lights definition Objects and apply them to the AtriumFirstThreeFloors Space Type as shown in Figs. 8.18 and 8.19. This will apply a high efficiency lighting level to the atrium along with a typical operation schedule.

Navigate to the Schedules (■) Tab to examine the lighting Schedules we just applied (Fig. 8.20).

Before running the simulation, navigate to the Variables (■) Tab and turn on output variable requests for "Zone Lights Electric Energy" and "Daylighting Reference Point 1 Illuminance" as shown in Fig. 8.21. This will allow us to see the electric energy used for lighting in each Thermal Zone to verify that the Daylighting Controls Object is indeed reducing lighting energy in response to available daylight.

Now we are ready to run our first daylighting simulation using EnergyPlus as the daylight calculation engine. Navigate to the Run Simulations (■) Tab and run the simulation as shown in Fig. 8.22. Note the additional messages in the run log related to daylighting.

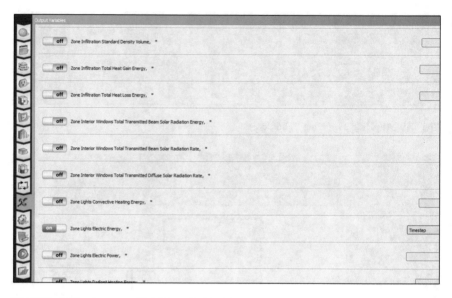

**Fig. 8.21**  Adding an output Variable to the Model

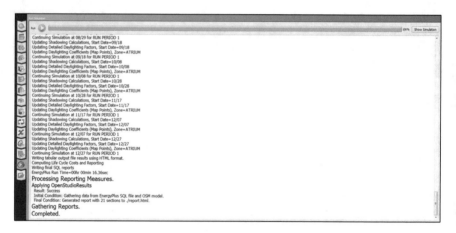

**Fig. 8.22**  Running a simulation using EnergyPlus for Daylighting calculations

The eplusout.sql file generated by the EnergyPlus simulation also includes Illuminance Map output. Illuminance Map vary throughout the day and throughout the year, so it's important to take that into consideration when reviewing Illuminance Map data. Figures 8.23 and 8.24 show two Illuminance Map plots for the atrium. Recall that the windows in the atrium face East. Notice how the illuminance levels in Fig. 8.23 are much higher in the in the morning due to direct sun exposure than they are later in the day in Fig. 8.24.[9] For reference,

---

[9] Note that EnergyPlus reports time as "standard" time. If daylight savings is in effect, then local time will be 1 h ahead of standard time.

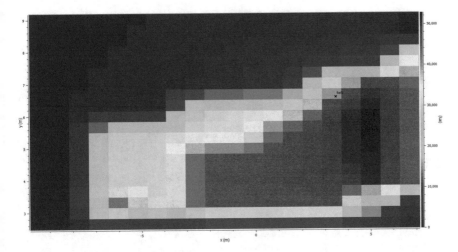

**Fig. 8.23**   EnergyPlus Illuminance Map on 6/21 at 8:00 AM DST

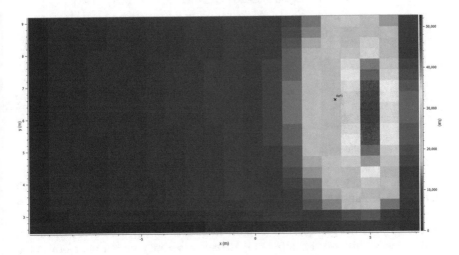

**Fig. 8.24**   EnergyPlus Illuminance Map on 6/21 at 1:00 PM DST

illuminance levels of 50,000 lux are much higher than needed for task lighting and will likely result in occupant discomfort.

We can examine the impact of daylighting controls on electric lighting by plotting the daylight illuminance calculated at the control point alongside the electric lighting energy in Fig. 8.25. Compare the lighting energy use profile to the lighting schedule in Fig. 8.20. Notice that during the morning when there is excessive daylight illumination the electric lights do not consume energy. Later in the day, as the sun moves to the other side of the building, daylight illumination levels drop, and the electric lights are turned back on to provide sufficient illumination for the remainder of the work day.

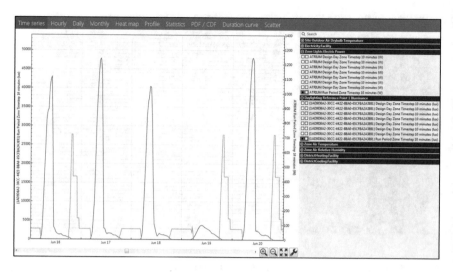

**Fig. 8.25** EnergyPlus Daylighting Illuminance and Electric Lighting Energy

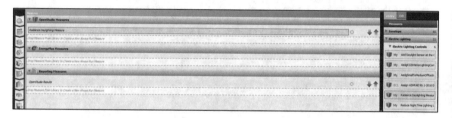

**Fig. 8.26** OpenStudio Radiance Measure added to workflow

At this point, we have successfully run a daylighting simulation using the split-flux calculation method available in EnergyPlus. EnergyPlus can provide a rough estimate of the energy savings potential for daylighting controls as well as alert us to potential glare issues with the large East facing windows. Depending on our needs, this may be sufficient. However, we may need to perform more accurate daylighting simulation to increase confidence in savings estimates, or to persuade an architect that a glare control feature is required in a design.

The only change necessary in our OpenStudio Model is the addition of the OpenStudio Radiance Measure. This measure can be found on the BCL under the "Electric Lighting Controls" category. Simply download this measure and drag it into your workflow on the Measures (▦) Tab as shown in Fig. 8.26.

The Radiance Measure has several parameters that control its operation as shown in Fig. 8.27. If checked, the Apply Schedules option applies lighting schedule reductions based on calculation of available daylight illuminance and removes

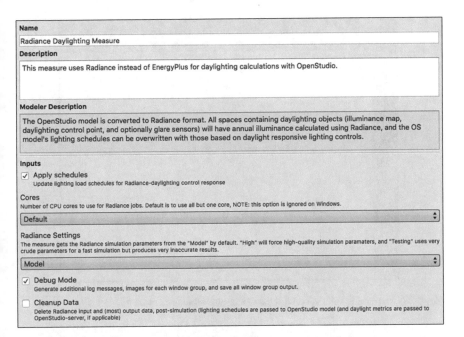

**Fig. 8.27**  OpenStudio Radiance measure parameters

Daylighting Control Objects in the Model prior to EnergyPlus simulation. The Cores input allows you to determine how many Radiance simulation processes may run in parallel during the Measure application. The Radiance Settings parameter allows the user to select between fast simulation settings for initial testing and debugging or higher quality settings for final calculations. The Debug and Cleanup Data settings control which intermediate Radiance files are retained after the Measure runs.

Once the Measure has been added and configured, simply return to the Run Simulations (▣) Tab and re-run your simulation as shown in Fig. 8.28. Note that the actual Radiance simulation occurs during processing of the OpenStudio Radiance Measure, so this measure will take more time to apply than most other measures.

After re-running the simulation, you can use ResultsViewer or DView to open eplusout.sql results from EnergyPlus as well as the radout.sql file generated by the Radiance Measure (this file is located in the \run\000_RadianceMeasure\ radiance\output directory). Open the radout.sql file to see Illuminance Map data generated by the OpenStudio Radiance Measure. Compare the results calculated with Radiance in Figs. 8.29, 8.30, and 8.31 with those calculated by EnergyPlus in Figs. 8.23, 8.24, and 8.25. What differences do you see?

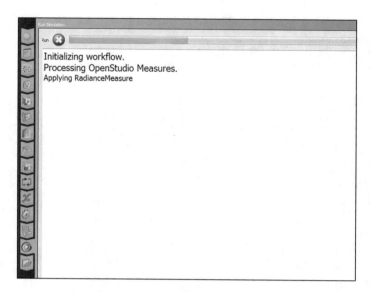

**Fig. 8.28** Re-running simulation with the OpenStudio Radiance measure

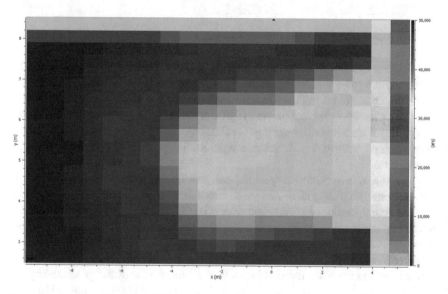

**Fig. 8.29** Radiance Illuminance Map on 6/21 at 8:00 AM DST

**Fig. 8.30**  Radiance Illuminance Map on 6/21 at 1:00 PM DST

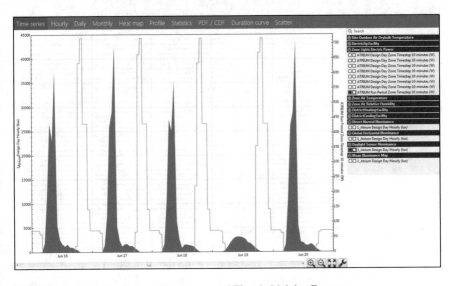

**Fig. 8.31**  Radiance Daylighting Illuminance and Electric Lighting Energy

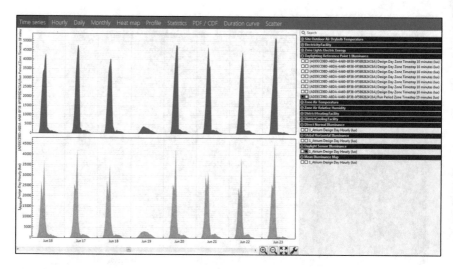

**Fig. 8.32** Comparison of EnergyPlus and Radiance Illuminance Calculations

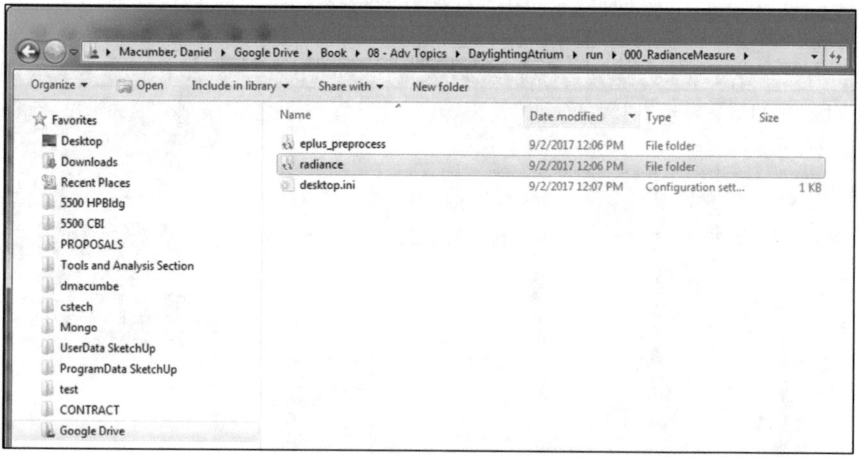

**Fig. 8.33** OpenStudio Radiance Measure run directory

DView can also be used to directly compare the daylight illuminance calculated at the Daylighting Control Object by EnergyPlus to that calculated by Radiance as shown in Fig. 8.32.

If you configured the Radiance Measure as shown in Fig. 8.27, several useful intermediate files generated by the Measure will have been preserved in the Measure's run directory shown in Fig. 8.33.

One such file is a scene rendering, which is generated at each Daylighting Control point as shown in Fig. 8.34. Scene renderings are produced at the position of the Daylighting Control looking upwards. They allow you to visualize what the Daylighting Control effectively observes at various times in the simulation.

**Fig. 8.34** Radiance scene rendering

| | A | B | C | D | E | F | G |
|---|---|---|---|---|---|---|---|
| 1 | # OpenStudio Daylight Metrics Report | | | | | | |
| 2 | # Building average daylight metrics (daylit spaces): DA = 0.46 cDA = 0.63 UDI = 0.57 | | | | | | |
| 3 | #[space_name] | [metric(setpoint)] | [input_hours_range] | [metric_value] | [hours_met] | [input_hours] | |
| 4 | 1_Atrium | DA(500) | Daylit Hours | 0.46 | 2158 | 4710 | |
| 5 | 1_Atrium | conDA(500) | Daylit Hours | 0.63 | 2978 | 4710 | |
| 6 | 1_Atrium | UDI(100-3000) | Daylit Hours | 0.57 | 2668 | 4710 | |
| 7 | 1_Atrium | sDA(300) | 8AM-5PM (10 hours/day per IESNA LM-83-12) | 0.57 | 2087 | 3650 | |
| 8 | | | | | | | |

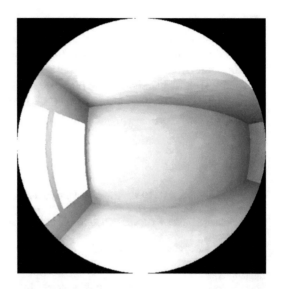

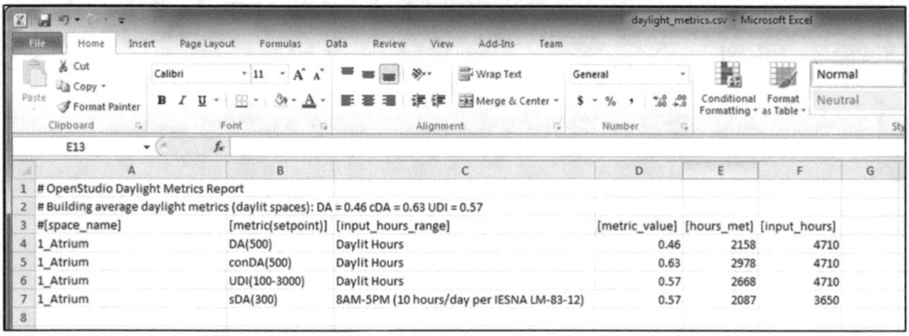

**Fig. 8.35** Additional Daylighting Metrics

Finally, the OpenStudio Radiance Measure calculates additional daylighting metrics as shown in Fig. 8.35.

This is but one advanced application of OpenStudio that makes use of a separate engine. In the next Chapter, we will delve more deeply into OpenStudio Measures, the fundamental OpenStudio building block that enables such applications to be built and integrated into sophisticated analysis workflows.

## 8.4 Additional Exercises

The IESNA Lighting Handbook provides recommended illuminance levels for a variety of space types. The recommended illuminance level for a "Lobby – Office/ General" space type is 200–300 lux. Results in this exercise show that direct illuminance levels in the atrium space can reach up to 50,000 lux at the sensor location. Check other points in the space to see what illuminance levels are right by the window, in the far corner, or in the middle of the room. What happens if the building is rotated so the atrium windows face North, South, or West? Try changing the "Visible Transmittance" of the "Simple Glazing System Window Materials" on the Materials Sub-Tab, what impact does that have?

# References

Ward G, Shakespeare R (1998) Rendering with radiance. Morgan Kaufmann, Waltham
https://bcl.nrel.gov/node/39783
http://bigladdersoftware.com/epx/docs/8-7/engineering-reference/time-step-daylighting-calcula-tion.html
https://cdn2.hubspot.net/hub/155785/file-18058478-pdf/docs/daylighting_research_-_us_govern-ment_report.pdf
https://gbxml.org
https://github.com/GreenBuildingXML/Sample-gbXML-Files
https://nrel.github.io/OpenStudio-user-documentation/reference/sketchup_plugin_interface
http://www.sciencedirect.com/science/article/pii/S0378778806000624
https://sketchup.com.

# Chapter 9
# The OpenStudio Software Development Kit

## 9.1 Introduction

As discussed in Chap. 1, OpenStudio is not a single energy modeling tool. Rather, it is an SDK or platform, designed to reduce the cost and time to create a variety of energy efficiency assessment applications. The OpenStudio Application and PAT, presented in previous chapters, are intended as examples of using the SDK to create software in C++ and Electron/Angular respectively. A third example of creating new functionality with the SDK is the OpenStudio Measure introduced in Chap. 6. OpenStudio Measures are the most accessible means of creating new capability with OpenStudio and represent the "gateway" to more advanced application development. For this reason, the bulk of Chap. 9 is devoted to adapting existing Measures or creating new ones to add functionality to the OpenStudio Application or PAT.

## 9.2 OpenStudio Measure Overview

We have already explored the basic concept of an OpenStudio Measure along with the BCL as a convenient source for them in Chap. 6. In Chap. 7, we extended our understanding of OpenStudio Measures as a means of optimizing building energy performance and calibrating models against consumption data. Chapter 8 highlighted an advanced use of Measure scripting to combine Radiance and EnergyPlus analyses for more accurate daylighting design savings estimates. Another example

---

The original version of this chapter was revised. A correction to this chapter can be found at https://doi.org/10.1007/978-3-319-77809-9_10

**Electronic Supplementary Material:** The online version of this chapter (https://doi.org/10.1007/978-3-319-77809-9_9) contains supplementary material, which is available to authorized users.

© Springer International Publishing AG, part of Springer Nature 2018
L. Brackney et al., *Building Energy Modeling with OpenStudio*,
https://doi.org/10.1007/978-3-319-77809-9_9

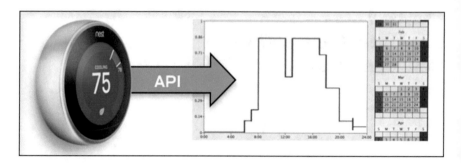

**Fig. 9.1**   NEST thermostat Measure concept

integrates[1] OpenStudio models with the GLHEPro[2] Ground Loop Heat Exchanger
Design tool. Lastly, the authors know of one enterprising OpenStudio user who wrote
a Measure to access measured occupancy data from his Nest thermostat as a means
of creating Model schedules (Fig. 9.1). We now turn our attention to how Measures
actually work and how to extend OpenStudio's functionality by creating new ones.

OpenStudio Measures are ZIP archives comprised of <u>at least</u> two files: measure.xml
and measure.rb. The measure.xml file is an eXtensible Markup Language (XML) file
containing descriptive information (metadata) about the measure. Measure metadata is
used by the BCL and OpenStudio applications to identify where and how a measure
might be used, what arguments the measure may accept, when it was last modified, etc.
The measure.xml file, while written in plain text, can be difficult to read and edit. XML
editor/viewers are widely available[3] and helpful in working with Measure metadata.
Figure 9.2 illustrates codebeautify.org's web-based XML viewer inspecting a raw
measure.xml (on left) and a convenient tree view of the same content (on right).

Readily apparent in the tree view are the Measure's name, when it was last modified, a high level description, a more detailed explanation of how the Measure
works, its arguments, search tags, and more. In this case, the "*set_gas_burner_effi-
ciency*" Measure contains ten arguments, many of which are optional. The first two
named "*object*" and "*eff*" are required and tell the Measure which of the Model's air
loops it should operate on, the default being all of them, and the fractional burner
efficiency, which defaults to 0.95. Argument names, type definitions, etc. are of
particular significance in describing how information is passed between OpenStudio
applications and the measure.rb code itself.

If the OpenStudio Measure metadata standard defines how it can be connected
with OpenStudio applications, much like a puzzle piece, then the measure.rb represents the picture on the puzzle piece. Measures may contain additional files such as
design documents, test cases, and more as part of the Measure's "payload," but the
measure.rb contains the actual Ruby code that is executed whenever the measure is
invoked by an application. Figure 9.3 illustrates the overall "anatomy" of a Measure
as it might appear on the BCL.

---

[1] https://bcl.nrel.gov/search/site/GLHEPro?f[0]=bundle%3Anrel_measure
[2] https://hvac.okstate.edu/glhepro/overview
[3] http://codebeautify.org/xmlviewer

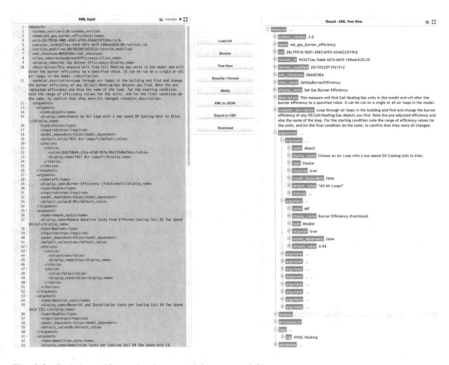

**Fig. 9.2** Code beautify used to inspect a Measure.xml file

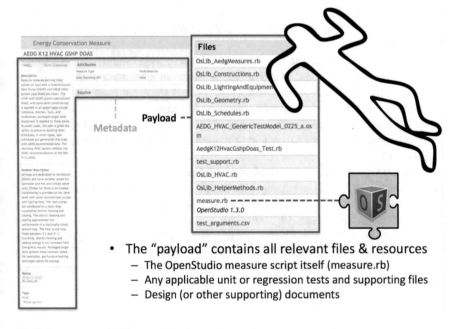

- The "payload" contains all relevant files & resources
  - The OpenStudio measure script itself (measure.rb)
  - Any applicable unit or regression tests and supporting files
  - Design (or other supporting) documents

**Fig. 9.3** Anatomy of a Measure

The following "snippet" of Measure Ruby code provides some flavor for what Measure scripts look like. In this example, the Measure loops across all surfaces in the Model. If a surface is an exterior wall, then a new construction is applied to that surface.

```
# Measure structure to replace exterior wall orig construction with new
model.getSurfaces.each do |s|
    if s.outsideBoundaryCondition == "Outdoors" and s.surfaceType == "Wall"
        if s.construction.name.get == orig
            s.setConstruction(new)
```

Sections 9.3 and 9.4 cover the rudiments of modifying existing measure.xml and measure.rb files or creating new ones. Section 9.5 presents the important topic of Measure testing, and additional files one might want to include as part of a Measure's payload to continuously verify that it works as intended. The reader is referred to the Measure Writer's Reference Guide on the OpenStudio website[4] as an additional resource for use alongside these sections of the text.

## 9.3   Adapting an Existing Measure

Recall from Chap. 6 that Measures frequently start out on the BCL and are then downloaded to the user's local library. In the case of PAT, Measures may then be added from that local library to a specific project. Figure 9.4 shows PAT's now familiar Measure management window with a range of fenestration Measures. Two of these have already been downloaded and are available to add to this PAT project.

Note that once a Measure has been downloaded, its ⊕ Button changes to a 🔧 Button. Clicking 🔧 creates a copy of the measure in a separate directory designated as "MyMeasures." This directory is distinct from the local library and is intended for use in creating new or customized Measures. Figure 9.5 illustrates a "flow" of Measures from the BCL to a user's local Measures directory, which can then be used directly in a PAT project or copied into MyMeasures for modification and use in PAT.

Both the OpenStudio Application and PAT have a Preferences Menu option that allows the user to specify (or change) the location of this directory. Measures in the MyMeasures directory will be denoted with a "My" label in PAT or an 🔧 icon in the OpenStudio Application (Fig. 9.6).

Highlighting a Measure in the OpenStudio Application also enables a copy Measure option indicated by a 🔲 Button. This behaves identically to PAT's 🔧 Button and makes a copy of the Measure XML and .rb files to your MyMeasures directory for further editing (Fig. 9.7).

---

[4] http://nrel.github.io/OpenStudio-user-documentation/reference/measure_writing_guide/

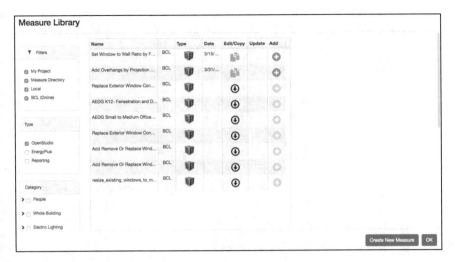

**Fig. 9.4** Measure library with BCL fenestration content shown

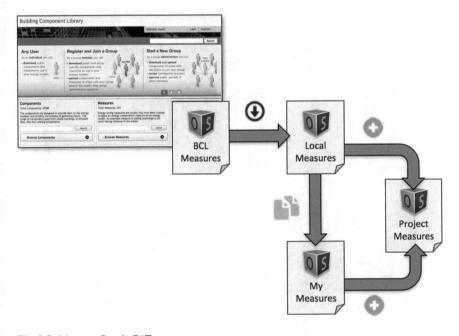

**Fig. 9.5** Measure flow in PAT

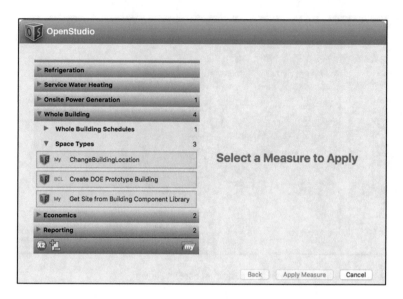

**Fig. 9.6**   OpenStudio Apply Measure Now dialog highlighting BCL and MyMeasures content

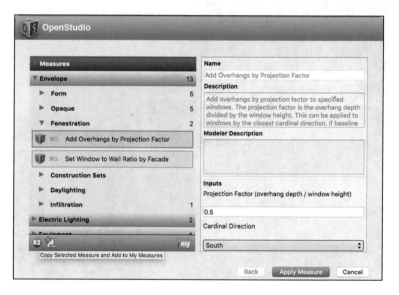

**Fig. 9.7**   Preparing to make a copy of a Measure in the OpenStudio Application

## 9.4   Checkpoint Thirteen: Adapting Measures

One of the best ways to learn how to write Measures is to modify an existing Measure. By taking this approach, users who are not programming experts can gain the confidence and skills to eventually develop their own Measures from scratch. In this exercise, we will copy the measure named "GasHeatingCoilEfficiency" and modify it to set the efficiency of gas fired hot water boilers instead.

### 9.4.1   Text Editor

For this exercise, you will need a text editor. Although a basic text editor like Notepad for Windows will work, a text editor with syntax highlighting for the Ruby programming language makes the process much easier. There are many programs available online. Notepad++ is a good free option for Windows, and Mac users might consider TextMate. The authors would not presume to suggest anything for Linux users, lest we provoke a VI/Emacs altercation.

### 9.4.2   Programming Background

Writing Measures involves basic computer programming with the Ruby scripting language. If basic programming concepts are unfamiliar to you, we recommend beginning with an online tutorial.[5] Spending an hour or two going through one of the many excellent online Ruby tutorials will be well worth the effort. Here are few key concepts you will need to comprehend, edit, or write most measures:

- Data types (String, Double, Integer, Array, Boolean),
- Variables,
- If & Case Statements, and
- For & For Each Loops.

### 9.4.3   Copy the Measure

The first step in adapting a Measure for a new purpose is to find an existing one that has similar functionality to what you are trying to achieve. For example, if your goal were to modify wall insulation, a Measure that modified roof insulation might be a good candidate. For the purpose of this exercise, the Measure to be copied will be "GasHeatingCoilEfficiency."

---

[5] https://www.ruby-lang.org/en/documentation/quickstart/

1. Open the *PrimarySchoolHVAC.osm* Model from Chap. 4 in the OpenStudio Application.
2. Open the "Apply Measure Now" dialog under "Components & Measures"
3. Find the "Gas Heating Coil Efficiency" Measure under HVAC > Heating
4. Make a copy of the Measure using the ▣ Button.
5. Change the name to: Gas Fired Boiler Efficiency.

### 9.4.4   Review the Measure

After the new Measure is created, a folder will be opened that contains a/*tests* directory, a *measure.xml file,* and the new *measure.rb* file. Open the *measure.rb* file in your text editor. It should look something like Fig. 9.8 below.

The first step is to get a general idea of how the existing code works. Well-written measures should have comments describing what the code is doing at each step. These comments are denoted by the # symbol at the beginning of the line. Read these comments in order, starting at the top of the file and moving down to the bottom. Don't worry about the rest of the code in the file. You may need to repeat this process a few times to get the general idea. At the end of this process, you should have a general idea of the steps that are involved in the measure. If a measure is insufficiently commented or is very complex, it is not a good starting point. Rather than wasting time learning from a difficult example, look for a simpler measure to start with.

With the general idea in mind, start again at the first comment. Read the code immediately below this comment. Use your basic programming knowledge and energy modeling expertise to understand what is going on. You can add additional comments to the file on a line-by-line basis can help you document the progress you are making and keep the ideas straight. This process sounds tedious, but after doing it once or twice you will have a much better understanding of how the measure works.

```
# Start the measure
class GasBoilerEfficiency < OpenStudio::Ruleset::ModelUserScript

  # Define the name that a user will see
  def name
    return "Gas Boiler Efficiency"
  end

  # Define the arguments that the user will input
  def arguments(model)
    args = OpenStudio::Ruleset::OSArgumentVector.new

    # Make an argument for the COP
    eff = OpenStudio::Ruleset::OSArgument::makeDoubleArgument("eff",true)
    eff.setDisplayName("Rated Efficiency")
    eff.setDescription("Rated gas burner efficiency of the gas heating coil")
    eff.setUnits("%")
    eff.setDefaultValue(0.8)
    args << eff

    return args
  end
```

**Fig. 9.8**  Reading through a Measure

### 9.4.5   Modify the Measure

After gaining an understanding of the code, we are going to make our first edit to the file. In this case, we are simply changing the description and default value for the measure argument from gas burner efficiency to boiler efficiency. It is a good idea to start small, make minor changes, and then test before moving on. Find the section of the measures shown in Fig. 9.9 below, make the changes specified, and save the file.

Before making any other changes, run the measure using "Apply Measure Now" and review the resulting Model. Ensure that the changes you made are actually reflected in the Model. In this case, the changes to the argument can be seen in the dialog, as shown in Fig. 9.10.

If the measure didn't run, go back and correct any typos or other errors you may have made. Generally, the line number for the error will be shown in the message. For example, Fig. 9.11 identifies the location of a typo.

Proceed in this fashion, making small changes and testing, until the measure is to your liking. Until you have performed this process a few times, resist the temptation to change multiple things at once. Think of the process as an experiment; if you change two variables and the measure doesn't run, you won't know which change was responsible for the failure.

Next, modify the allowable boiler efficiencies. Change the code as shown in Fig. 9.12 so that the boiler efficiency must be between 0.65 (65%) and 1 (100%). This range is acceptable for our purposes, but be aware that an existing building could contain an old boiler in disrepair that falls outside of these bounds these bounds.

Next, modify the code to change boilers instead of gas heating coils, as shown in Fig. 9.13.

Next, modify the code that reports whether or not the Measure was applicable, as shown in Fig. 9.14.

Next, modify the code that reports the initial and final condition of the Model, as shown in Fig. 9.15.

```
Change this:

    # Make an argument for the efficiency
    eff = OpenStudio::Ruleset::OSArgument::makeDoubleArgument("eff",true)
    eff.setDisplayName("Rated Efficiency")
    eff.setDescription("Rated gas burner efficiency of the gas heating coil")
    eff.setUnits("%")
    eff.setDefaultValue(0.8)
    args << eff

To this:

    # Make an argument for the efficiency
    eff = OpenStudio::Ruleset::OSArgument::makeDoubleArgument("eff",true)
    eff.setDisplayName("Rated Efficiency")
    eff.setDescription("Rated efficiency of the boiler")
    eff.setUnits("%")
    eff.setDefaultValue(0.85)
    args << eff
```

**Fig. 9.9**  Initial edits to the new Measure

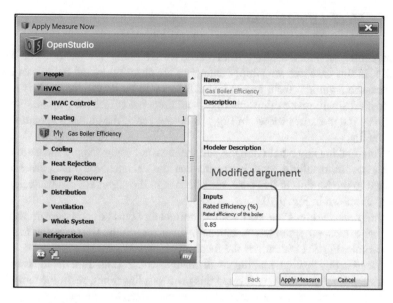

**Fig. 9.10**  Verifying that the argument changes were made correctly

**Fig. 9.11**  Identifying the location of a Measure error

```
Change this:

    # Check the efficiency for reasonableness
    # Efficiency must be between zero and one
    if eff <= 0 || eff >= 1
      runner.registerError("Efficiency must be between zero and one.  You entered #{eff}.")
      return false
    end

To this:

    # Check the efficiency for reasonableness
    # Efficiency must be between 0.65 and 1
    if eff <= 0.65 || eff >= 1
      runner.registerError("Efficiency must be between 0.65 and 1.  You entered #{eff}.")
      return false
    end
```

**Fig. 9.12**  Making a second change to the Measure

```
Change this:

    # Loop through the gas heating coils in the model
    num_htg_coils_changed = 0
    orig_effs = []
    model.getCoilHeatingGass.each do |htg_coil|

        # Get the original efficiency and store it
        eff_original = htg_coil.gasBurnerEfficiency
        orig_effs << eff_original

        # Change the efficiency to the new value
        htg_coil.setGasBurnerEfficiency(eff)

        # Report the change of efficiency
        runner.registerInfo("Changing the efficiency of #{htg_coil.name} from #{eff_original} to
#{eff} ")

        # Add to the number of coils changed
        num_htg_coils_changed += 1

    end

To this:

    # Loop through the boilers in the model
    num_blrs_changed = 0
    orig_effs = []
    model.getBoilerHotWaters.each do |boiler|

        # Get the original efficiency and store it
        eff_original = boiler.nominalThermalEfficiency
        orig_effs << eff_original

        # Change the efficiency to the new value
        boiler.setNominalThermalEfficiency(eff)

        # Report the change of efficiency
        runner.registerInfo("Changing the efficiency of #{boiler.name} from #{eff_original} to
#{eff} ")

        # Add to the number of boilers changed
        num_blrs_changed += 1

    end
```

**Fig. 9.13** Making a third change to the Measure

```
Change this:

    # Not applicable if no coils were changed
    if num_htg_coils_changed == 0
        runner.registerAsNotApplicable("Not applicable; model contains no gas heating coils.")
        return true
    end

To this:

    # Not applicable if no boilers were changed
    if num_blrs_changed == 0
        runner.registerAsNotApplicable("Not applicable; model contains no boilers.")
        return true
    end
```

**Fig. 9.14** Making a fourth Measure change

```
Change this:

    # Record the initial efficiency range of the heating coils
    runner.registerInitialCondition("Gas coils had efficiencies between #{orig_effs.min * 100}% to
#{orig_effs.max * 100}%.")

    # Record the final efficiency of the heating coils
    runner.registerFinalCondition("#{num_htg_coils_changed} gas coils were set to #{eff * 100}%
efficiency.")

To this:

    # Record the initial efficiency range of the boilers
    runner.registerInitialCondition("Boilers had efficiencies between #{orig_effs.min * 100}% to
#{orig_effs.max * 100}%.")

    # Record the final efficiency of the boilers
    runner.registerFinalCondition("#{num_blrs_changed} boilers were set to #{eff * 100}%
efficiency.")
```

**Fig. 9.15** Final Measure change

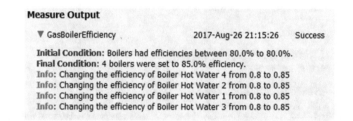

**Fig. 9.16** Output from a successful application of our Measure

The final step is testing the Measure on a Model. Run the Measure via "Apply Measure Now" on our school Model. The result should be a dialog that shows the newly modified messages, which refer to boilers instead of gas heating coils, and show the newly modified default efficiency, as shown in Fig. 9.16.

If the Measure output looks good, the next step is to verify the Object values in the Model itself. Click [Accept Changes] in the "Apply Measure Now" dialog. Proceed to the HVAC (■) Tab, and then navigate to the Hot Water Loop. Click on the boiler Object and ensure that the efficiency value matches expectations, as shown in Fig. 9.17. If it does not, use the "File/Revert to Saved" menu option to discard these changes and reload the Model as it was before the Measure was applied.

## 9.5   Creating a New Measure

Both the OpenStudio Application and PAT enable the creation of new Measures from scratch. Clicking the [Create New Measure] Button in PAT or the ■ Button in the OpenStudio Application opens a window to capture information about the new Measure (Fig. 9.18). The astute reader will quickly realize that the contents of this dialog are used to create a new measure.xml file, and allow the author to specify the nature of the Measure, its intended use, and how it should show up within a Measure search.

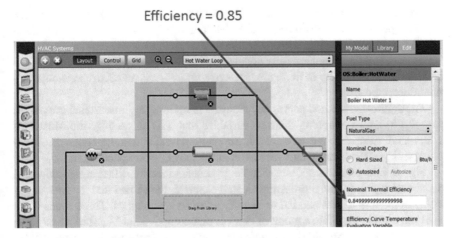

**Fig. 9.17**   Boiler efficiency modified successfully by our new Measure

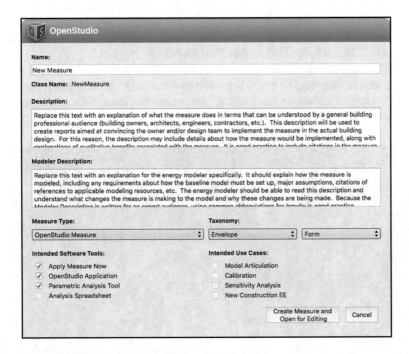

**Fig. 9.18**   OpenStudio Application Measure creation dialog

The Measure creation process also generates a measure.rb template file, which can be opened in a text editor or a Ruby Integrated Development Environment (IDE) like RubyMine (Fig. 9.19). The template file includes placeholders or "stubs" for required methods that are fully documented in the OpenStudio Measure Writer's Reference Guide. Many of these are self-explanatory – e.g. name, description, modeler_description, etc.

The arguments method parses any required or optional arguments that are passed into the Measure, mapping them to variables that will be used by the Measure's code. As illustrated above, new Measure contains a brief code snippet that illustrates using a single string argument with a default value of "New space name."

Figure 9.20 presents another sample arguments method, illustrating additional input data types. In this more sophisticated example, the Model that the Measure is being applied to is first queried to identify all of its space types. These space types are then used to populate a menu of choices that comprise the Measure's first argument dynamically. An additional (default) option "*Entire Building*" is also added, allowing the user to specify that the Measure should act on each space in the Model. A second floating point numerical argument related to "power reduction" is also expected by the Measure and defaults to 30. The reader is referred to the Measure Writing Guide for complete documentation on Measure arguments. Existing Measures in the BCL are also an excellent source of examples.

A Measure's run method is where the "real work" is performed. Logic to ensure that inputs are meaningful, perform transformations to the Model, and reportstatus are all contained within the run method. The run method created for a new Measure is shown in Fig. 9.21.

```
measure.rb
1    # see the URL below for information on how to write OpenStudio measures
2    # http://nrel.github.io/OpenStudio-user-documentation/reference/measure_writing_guide/
3
4    # start the measure
5    class NewMeasure < OpenStudio::Ruleset::ModelUserScript
6
7      # human readable name
8      def name
9        return "New Measure"
10     end
11
12     # human readable description
13     def description
14       return "Replace this text with an explanation of what the measure does in terms that can be understood by a general building profes
15     end
16
17     # human readable description of modeling approach
18     def modeler_description
19       return "Replace this text with an explanation for the energy modeler specifically.  It should explain how the measure is modeled,
20     end
21
22     # define the arguments that the user will input
23     def arguments(model)
24       args = OpenStudio::Ruleset::OSArgumentVector.new
25
26       # the name of the space to add to the model
27       space_name = OpenStudio::Ruleset::OSArgument.makeStringArgument("space_name", true)
28       space_name.setDisplayName("New space name")
29       space_name.setDescription("This name will be used as the name of the new space.")
30       args << space_name
31
32       return args
33     end
34
35     # define what happens when the measure is run
36     def run(model, runner, user_arguments)
37       super(model, runner, user_arguments)
38
39       # use the built-in error checking
40       if !runner.validateUserArguments(arguments(model), user_arguments)
41         return false
42       end
43
44       # assign the user inputs to variables
45       space_name = runner.getStringArgumentValue("space_name", user_arguments)
```

**Fig. 9.19** New Measure template file edited within the RubyMine IDE

```ruby
#define the arguments that the user will input
def arguments(model)
  args = OpenStudio::Ruleset::OSArgumentVector.new

  #make a choice argument for model objects
  space_type_handles = OpenStudio::StringVector.new
  space_type_display_names = OpenStudio::StringVector.new

  #putting model object and names into hash
  space_type_args = model.getSpaceTypes
  space_type_args_hash = {}
  space_type_args.each do |space_type_arg|
    space_type_args_hash[space_type_arg.name.to_s] = space_type_arg
  end

  #looping through sorted hash of model objects
  space_type_args_hash.sort.map do |key,value|
    #only include if space type is used in the model
    if value.spaces.size > 0
      space_type_handles << value.handle.to_s
      space_type_display_names << key
    end
  end

  #add building to string vector with space type
  building = model.getBuilding
  space_type_handles << building.handle.to_s
  space_type_display_names << "*Entire Building*"

  #make a choice argument for space type
  space_type = OpenStudio::Ruleset::OSArgument::makeChoiceArgument("space_type", space_type_handles, space_type_display_names)
  space_type.setDisplayName("Apply the Measure to a Specific Space Type or to the Entire Model.")
  space_type.setDefaultValue("*Entire Building*") #if no space type is chosen this will run on the entire building
  args << space_type

  #make an argument for reduction percentage
  power_reduction_percent = OpenStudio::Ruleset::OSArgument::makeDoubleArgument("power_reduction_percent",true)
  power_reduction_percent.setDisplayName("Power Reduction (%).")
  power_reduction_percent.setDefaultValue(30.0)
  args << power_reduction_percent

  return args
end #end the arguments method
```

**Fig. 9.20** Example Measure arguments method

```ruby
# define what happens when the measure is run
def run(model, runner, user_arguments)
  super(model, runner, user_arguments)

  # use the built-in error checking
  if !runner.validateUserArguments(arguments(model), user_arguments)
    return false
  end

  # assign the user inputs to variables
  space_name = runner.getStringArgumentValue("space_name", user_arguments)

  # check the space_name for reasonableness
  if space_name.empty?
    runner.registerError("Empty space name was entered.")
    return false
  end

  # report initial condition of model
  runner.registerInitialCondition("The building started with #{model.getSpaces.size} spaces.")

  # add a new space to the model
  new_space = OpenStudio::Model::Space.new(model)
  new_space.setName(space_name)

  # echo the new space's name back to the user
  runner.registerInfo("Space #{new_space.name} was added.")

  # report final condition of model
  runner.registerFinalCondition("The building finished with #{model.getSpaces.size} spaces.")

  return true

end
end
```

**Fig. 9.21** Run method from the new Measure template

This run method includes a call to OpenStudio's "runner.validateUserArguments" method that verifies that the user entered the required inputs with the correct data types. A Measure-specific check ensures that a non-empty space name was entered before proceeding. Measure authors are encouraged to qualify user inputs rigorously and invoke the "runner.registerError" method with a meaningful error message prior to executing "return false" code. This helps ensure that the Measure doesn't generate nonsensical models, providing meaningful feedback to the user.

Related to the topic of providing useful feedback to the user are a trio of OpenStudio methods; "runner.registerInitialCondition," "runner.registerInfo," and "runner.registerFinalCondition." We first experienced the products of these methods in Sect. 6.3.1 with an example Measure that reported the initial and final states of the Model when the ERV Measure was applied to a Model. The new Measure template reports the number of spaces in the initial Model, the name of the new space that was added, and the number of spaces in the final Model. The Measure's logic concludes with "return true," informing the program that called the Measure that it completed normally.

## 9.6   Checkpoint Fourteen: Creating a New Measure

In this exercise, we will create a new Measure from scratch, starting with just the basic template created by OpenStudio. This final activity assumes that you have read the OpenStudio Measure Writer's Reference Guide online. You will likely find the need to refer back to it often to understand the code that is being used.

The first step in writing a Measure is writing an outline of what it will do. This step may seem silly, but having an outline to refer back to when you get lost in the details of coding can be very helpful. For this example, we are going to make a Measure that adds a user-specified process load, of a user-specified fuel type to the largest Space in the Model. An outline of the steps includes:

1. Get user input for load type (electric or gas),
2. Get user input for the amount of load (W),
3. Identify the largest Space in the Model,
4. Create a process load of the appropriate type and wattage,
5. Assign it to the identified Space, and
6. Report out the load amount that was added and the associated Space.

The next step in the process is to go through the OpenStudio GUI and walk through the outline steps. Expand upon the outline by writing down the OpenStudio Object types to be used. This will make it easier to look up the documentation for these objects while writing the Measure. A revised outline might look something like this:

1. Get user input for load type (electric or gas),

   • *Choice input (dropdown list)*

2. Get user input for the amount of load (W),

   • *Number input*

3. Identify the largest Space in the Model,

   • *OpenStudio Space objects*

4. Create a process load of the appropriate type and wattage,

   • *If electric, use OpenStudio Electric Equipment Definition and Electric Equipment objects*
   • *If gas, use OpenStudio Gas Equipment Definition and Gas Equipment objects*

5. Assign it to the identified Space, and

   • *OpenStudio Space objects*

6. Report out the amount of load that was added, and the associated Space.

Create a new Measure using the OpenStudio Application, as shown in Fig. 9.18. Use the inputs shown in Fig. 9.22.

Open the new measure.rb file in a text editor. Select this Measure in the "Apply Measure Now" dialog. It will be found under Equipment > Electric Equipment. You should see that default arguments match those in the measure.rb file. They must be

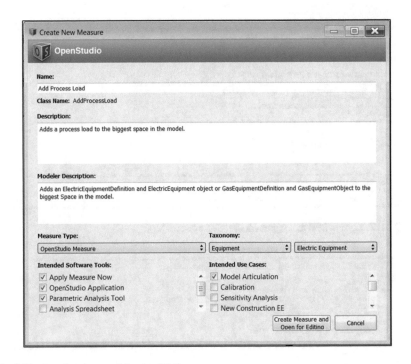

**Fig. 9.22**  Creating the new Measure XML

customized for our application. Create two Measure arguments: one for the load type and another for the wattage as shown in Fig. 9.23.

Add the code from Fig. 9.24 to assign the argument values to variables.

Figure 9.25 contains a code snippet that identifies the largest Space in a Model.

It is best practice to include error handling code and appropriate reporting in a Measure. Figure 9.26 is a simple example that reports if the Model contains no Spaces and exits. If Spaces do exist, then the Measure reports its name and size in ft$^2$.

The logic shown in Fig. 9.27 adds the appropriate load Object depending upon the fuel type and assigns it to the appropriate Space.

The Measure concludes in Fig. 9.28 with code to report the name of the load that was added to the largest Space.

Test the Measure on an example Model and ensure that it works as intended. If you have written the Measure correctly, the output should look like Fig. 9.29.

You've now written an entire Measure that may be used in the OpenStudio Application, PAT, or other OpenStudio-based applications. By writing an outline, planning the approach, and then dividing code up into small, manageable pieces; the Measure writing process is made less daunting. What tedious modeling tasks do you want to automate next?

```
# Get user input for process load type
load_type_chs = OpenStudio::StringVector.new
load_type_chs << 'Electric'
load_type_chs << 'Natural Gas'
load_type = OpenStudio::Ruleset::OSArgument::makeChoiceArgument('load_type', load_type_chs, true)
load_type.setDisplayName('Load Type')
load_type.setDescription('The type of process load to add.')
load_type.setDefaultValue('Electric')
args << load_type

# Get user input for load wattage
load_wattage = OpenStudio::Ruleset::OSArgument::makeDoubleArgument('load_wattage', true)
load_wattage.setDisplayName('Load Wattage')
load_wattage.setUnits('W')
load_wattage.setDefaultValue(1000.0)
args << load_wattage
```

**Fig. 9.23** Measure argument code

```
# assign the user inputs to variables
load_type = runner.getStringArgumentValue("load_type", user_arguments)
load_wattage = runner.getDoubleArgumentValue("load_wattage", user_arguments)
```

**Fig. 9.24** Assigning Measure arguments to Measure variables

```
# Find the biggest space (by floor area) in the model
big_space = nil
biggest_area_m2 = 0
model.getSpaces.each do |space|
  # Go to the next space unless this one is bigger than previous spaces
  next if space.floorArea < biggest_area_m2
  # Record this as the biggest space
  big_space = space
  biggest_area_m2 = space.floorArea
end
```

**Fig. 9.25** Identifying the largest Space in a Model

```
# Not applicable if no spaces was found
if big_space.nil?
  runner.registerAsNotApplicable("No spaces were found in the model.")
  return true
end

# Record the biggest space
biggest_area_ft2 = OpenStudio.convert(biggest_area_m2, 'm^2', 'ft^2').get
runner.registerInfo("The biggest space is #{big_space.name}, with a floor area of
#{biggest_area_ft2} ft2")
```

**Fig. 9.26** Error trapping logic for models with no spaces

```
# Make the correct type of load definition
case load_type
when 'Electric'
  # Create load definition
  load_def = OpenStudio::Model::ElectricEquipmentDefinition.new(model)
  load_def_name = "#{load_wattage.round} W Electric Load Definition"
  load_def.setName(load_def_name)
  load_def.setDesignLevel(load_wattage)

  # Create load instance
  load = OpenStudio::Model::ElectricEquipment.new(load_def)
  load_name = "#{load_wattage.round} W Electric Load"
  load.setName(load_name)
  load.setMultiplier(1.0)

  # Assign to biggest space
  load.setSpace(big_space)
  runner.registerInfo("Added #{load_name} to #{big_space.name}.")

when 'Natural Gas'
  # Create load definition
  load_def = OpenStudio::Model::GasEquipmentDefinition.new(model)
  load_def_name = "#{load_wattage.round} W Gas Load Definition"
  load_def.setName(load_def_name)
  load_def.setDesignLevel(load_wattage)

  # Create load instance
  load = OpenStudio::Model::GasEquipment.new(load_def)
  load_name = "#{load_wattage.round} W Gas Load"
  load.setName(load_name)
  load.setMultiplier(1.0)

  # Assign to biggest space
  load.setSpace(big_space)
  runner.registerInfo("Added #{load.name} to #{big_space.name}.")

end
```

**Fig. 9.27** Ruby code that adds electric or gas equipment objects to the largest Space in the Model

```
# Record the final condition
runner.registerFinalCondition("Added one #{load.name} to #{big_space.name}.")

return true
```

**Fig. 9.28** Reporting the final state of the Model after the Measure completes

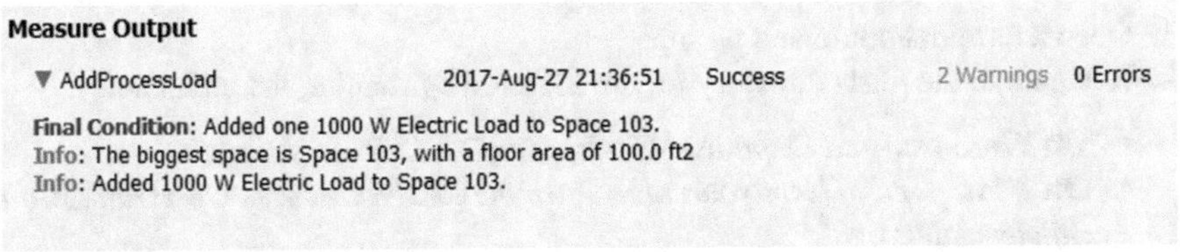

**Measure Output**

▼ AddProcessLoad            2017-Aug-27 21:36:51      Success              2 Warnings   0 Errors

**Final Condition:** Added one 1000 W Electric Load to Space 103.
**Info:** The biggest space is Space 103, with a floor area of 100.0 ft2
**Info:** Added 1000 W Electric Load to Space 103.

**Fig. 9.29** Successful addition of a process load to the largest Space in a Model

## 9.7   Checkpoint Fifteen: Testing Measures

As shown in the previous two exercises, the most obvious way to test that a Measure is working correctly is to apply the Measure to a Model and check that the output messages and resulting changes occurred as expected. Some best practices for this step include testing on more than one Model, where differences in the Model's contents may cause issues. Testing on a Model where the Measure should not be applicable is equally valuable to ensure that the applicability is reported correctly.

Advanced users with large collections of their own Measures, or companies where employees share common libraries of Measures need to ensure that Measures continue to work over time. In these situations, it is impractical to manually re-test Measure functionality each time a change is made to OpenStudio itself or Measure code.

Fortunately, OpenStudio borrows from software engineering best practices to provide a solution. This solution is known as unit testing. A unit test is simply a script that runs a piece of code and checks that certain conditions, known as assertions are met. Unit tests for OpenStudio Measures programmatically assert that a Measure behaves correctly for one or more test Models. Assertion is not a guarantee of consistent performance but is a strong indicator.

Navigate to your *MyMeasures* directory to locate your GasHeatingCoil Measure, then open the file called GasHeatingCoil_Test.rb from the /tests subfolder. Figure 9.30 below shows this unit test for the Gas Heating Coil Measure.

Following the approach in the previous checkpoint and read through the file to get a high level understanding of how it works. A quick read of the comments shows that this test:

1. Creates an instance of the Measure,
2. Creates a runner to run the Measure,
3. Loads the test Model,
4. Sets the arguments for the Measure,
5. Runs the Measure on the test Model using the supplied arguments,
6. Shows the output including information, warning, & error messages,
7. Asserts that the Measure was applied successfully,
8. Asserts that the gas heating coils in the Model have the expected efficiencies, and
9. Saves the modified Model for later inspection,

Next, run the unit test for this Measure by using the following steps:

1. Open a terminal/command prompt
2. Navigate to the /tests directory for the Measure by running the command:

   - **On Windows** – *cd C:\path\to\GasHeatingCoilEfficiency\tests*
   - **On Mac** – *cd /Users/yourname/OpenStudio/MyMeasures/GasHeatingCoil Efficiency\tests*

```ruby
require 'openstudio'
require 'openstudio/ruleset/ShowRunnerOutput'

require "#{File.dirname(__FILE__)}/../measure.rb"

require 'minitest/autorun'

class GasHeatingCoilEfficiency_Test < MiniTest::Unit::TestCase

  def test_change_efficiency
    # create an instance of the measure
    measure = GasHeatingCoilEfficiency.new

    # create an instance of a runner
    runner = OpenStudio::Measure::OSRunner.new(OpenStudio::WorkflowJSON.new)

    # load the test model
    translator = OpenStudio::OSVersion::VersionTranslator.new
    path = OpenStudio::Path.new(File.dirname(__FILE__) + "/MyPrimarySchoolHVACTest.osm")
    model = translator.loadModel(path)
    assert(model.is_initialized)
    model = model.get

    # get arguments
    arguments = measure.arguments(model)
    argument_map = OpenStudio::Ruleset.convertOSArgumentVectorToMap(arguments)

    # create hash of argument values.
    # If the argument has a default that you want to use, you don't need it in the hash
    args_hash = {}
    args_hash["eff"] = 0.86
    # using defaults values from measure.rb for other arguments

    # populate argument with specified hash value if specified
    arguments.each do |arg|
      temp_arg_var = arg.clone
      if args_hash.has_key?(arg.name)
        assert(temp_arg_var.setValue(args_hash[arg.name]))
      end
      argument_map[arg.name] = temp_arg_var
    end

    # run the measure
    measure.run(model, runner, argument_map)
    result = runner.result

    # show the output
    show_output(result)

    # assert that it ran correctly
    assert_equal("Success", result.value.valueName)

    # assert that all boilers in the model now have an efficiency of 0.86
    model.getCoilHeatingGass.each do |htg_coil|
      assert_equal(0.86, htg_coil.gasBurnerEfficiency, "Efficiency was not set correctly.")
    end

    # save the model to test output directory
    output_file_path = OpenStudio::Path.new(File.dirname(__FILE__) + "/output/test_output.osm")
    model.save(output_file_path, true)
  end

end
```

**Fig. 9.30**  A typical Measure unit test

3. Run the command:

- **On Windows** – C:\openstudio-2.3.0\bin\openstudio GasHeatingCoilEffici
  ency_Test.rb
- **On Mac** – /Applications/OpenStudio-2.3.0/bin/openstudio GasHeatingCoil
  Efficiency_Test.rb

The output in Fig. 9.31 shows the initial condition, the final condition, info, warning, and error messages. The last line shows that there were four assertions

```
Run options: --seed 63634

# Running tests:

[openstudio.measure.OSRunner] <1> Cannot find current Workflow Step
**MEASURE APPLICABILITY**
0 = Success
**INITIAL CONDITION**
Gas coils had efficiencies between 80.0% to 80.0%.
**FINAL CONDITION**
1 gas coils were set to 86.0% efficiency.
**INFO MESSAGES**
Changing the efficiency of Gas Htg Coil from 0.8 to 0.86
**WARNING MESSAGES**
**ERROR MESSAGES**
***Machine-Readable Attributes**
{
    "attributes": {
        "eff": 0.85999999999999999,
        "eff_display_name": "eff"
    },
    "openstudio_version": "2.2.0"
}



Finished tests in 1.343076s, 0.7446 tests/s, 2.9782 assertions/s.

1 tests, 4 assertions, 0 failures, 0 errors, 0 skips
```

**Fig. 9.31** Typical Measure unit test output

made, and that there were zero test failures. This means that all of the checks passed and that the Measure is asserted to work correctly.

Now, modify the unit test for the Gas Boiler Efficiency Measure that was created in the previous checkpoint. The modified code is shown on the next page with the changes highlighted (Fig. 9.32). Try to avoid looking at this code while making the modifications yourself and see how far you can get. Remember the strategy of making small changes and re-running. Remember that error messages typically show the line number where an error exists.

When you have successfully modified the unit test, the output will look like Fig. 9.33.

In order for unit tests to be useful, they must incorporate meaningful checks. By itself, checking that the Measure completed successfully is a poor assertion of success. The Measure could be setting values incorrectly or missing other key behaviors entirely. For this reason, best practice is to include all the checks you would perform when reviewing the output of a Measure successfully in the unit test. Obviously, writing these tests takes time, so the effort needs to be weighed against the cost. For those with large collections of Measures to manage, the upfront cost of developing these tests quickly pays for itself in avoided issues later on.

```
require 'openstudio'
require 'openstudio/ruleset/ShowRunnerOutput'

require "#{File.dirname(__FILE__)}/../measure.rb"

require 'minitest/autorun'

class GasBoilerEfficiency_Test < MiniTest::Unit::TestCase

  def test_change_efficiency
    # create an instance of the measure
    measure = GasBoilerEfficiency.new

    # create an instance of a runner
    runner = OpenStudio::Measure::OSRunner.new(OpenStudio::WorkflowJSON.new)

    # load the test model
    translator = OpenStudio::OSVersion::VersionTranslator.new
    path = OpenStudio::Path.new(File.dirname(__FILE__) + "/MyPrimarySchoolHVACTest.osm")
    model = translator.loadModel(path)
    assert(model.is_initialized)
    model = model.get

    # get arguments
    arguments = measure.arguments(model)
    argument_map = OpenStudio::Ruleset.convertOSArgumentVectorToMap(arguments)

    # create hash of argument values.
    # If the argument has a default that you want to use, you don't need it in the hash
    args_hash = {}
    args_hash["eff"] = 0.86
    # using defaults values from measure.rb for other arguments

    # populate argument with specified hash value if specified
    arguments.each do |arg|
      temp_arg_var = arg.clone
      if args_hash.has_key?(arg.name)
        assert(temp_arg_var.setValue(args_hash[arg.name]))
      end
      argument_map[arg.name] = temp_arg_var
    end

    # run the measure
    measure.run(model, runner, argument_map)
    result = runner.result

    # show the output
    show_output(result)

    # assert that it ran correctly
    assert_equal("Success", result.value.valueName)

    # assert that all boilers in the model now have an efficiency of 0.86
    model.getBoilerHotWaters.each do |boiler|
      assert_equal(0.86, boiler.nominalThermalEfficiency, "Efficiency was not set correctly.")
    end

    # save the model to test output directory
    output_file_path = OpenStudio::Path.new(File.dirname(__FILE__) + "/output/test_output.osm")
    model.save(output_file_path,true)
  end

end
```

**Fig. 9.32** Modified unit test for the Gas Boiler Efficiency Measure

## 9.8 The OpenStudio Command Line Interface

The OpenStudio CLI is a compact, cross-platform executable that processes OpenStudio "Workflow" (OSW) files[6] to create and simulate building energy models. Workflows chain together Model files (OSMs) and Measures to automate common modeling tasks. An OSW is a JSON (JavaScript Object Notation) file that

---

[6] https://nrel.github.io/OpenStudio-user-documentation/reference/command_line_interface/

```
Run options: --seed 16777

# Running tests:

[openstudio.measure.OSRunner] <1> Cannot find current Workflow Step
**MEASURE APPLICABILITY**
0 = Success
**INITIAL CONDITION**
Boilers had efficiencies between 80.0% to 80.0%.
**FINAL CONDITION**
4 boilers were set to 86.0% efficiency.
**INFO MESSAGES**
Changing the efficiency of Boiler Hot Water 4 from 0.8 to 0.86
Changing the efficiency of Boiler Hot Water 2 from 0.8 to 0.86
Changing the efficiency of Boiler Hot Water 1 from 0.8 to 0.86
Changing the efficiency of Boiler Hot Water 3 from 0.8 to 0.86
**WARNING MESSAGES**
**ERROR MESSAGES**
***Machine-Readable Attributes**
{
    "attributes": {
        "eff": 0.85999999999999999,
        "eff_display_name": "eff"
    },
    "openstudio_version": "2.2.0"
}



Finished tests in 1.477084s, 0.6770 tests/s, 4.7391 assertions/s.

1 tests, 7 assertions, 0 failures, 0 errors, 0 skips
```

**Fig. 9.33** Unit test output for Gas Boiler Efficiency Measure

completely specifies a modeling workflow including the initial (seed) Model, weather file, and specific Measures that are to be applied sequentially, along with any arguments the Measures may require (Fig. 9.34).[7]

If Measures are akin to ingredients, then an OSW is the recipe, and the CLI is the cook who uses both to create part of a tasty meal. Figure 9.35 illustrates the basic concept of using the CLI to execute a simple workflow.

As its name would suggest, the CLI is invoked from a Windows, Mac, or Linux command line prompt. Calling the openstudio executable with a -h or --help argument produces the information shown in Fig. 9.36.

Using the CLI, the implementation details of proposed building energy modeling application become software agnostic. Any programming language that can be used to create an OSW and call the CLI can utilize OpenStudio and its supported engines to perform detailed energy analysis.

---

[7] https://nrel.github.io/OpenStudio-user-documentation/reference/command_line_interface/#osw-structure

```
{
    "seed_file": "baseline.osm",
    "weather_file": "USA_CO_Golden-NREL.724666_TMY3.epw",
    "steps": [
        {
            "measure_dir_name": "IncreaseWallRValue",
            "arguments": {}
        },
        {
            "measure_dir_name": "IncreaseRoofRValue",
            "arguments": {
                "r_value": 45
            }
        },
        {
            "measure_dir_name": "SetEplusInfiltration",
            "arguments": {
                "flowPerZoneFloorArea": 10.76
            }
        },
        {
            "measure_dir_name": "DencityReports",
            "arguments": {
                "output_format": "CSV"
            }
        }
    ]
}
```

**Fig. 9.34** Example OSW file

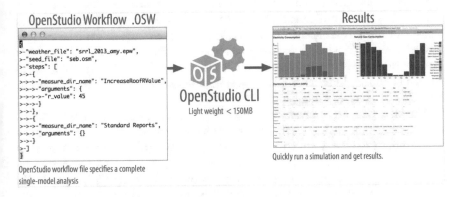

**Fig. 9.35** OpenStudio CLI enabled analysis (*Credit: Marjorie Schott*)

## 9.9   The OpenStudio Meta Command Line Interface

The CLI is a powerful tool for creating and simulating individual building models, but it requires a stream of OSW files to drive it. In Chap. 6, we saw this in action as we used PAT to manually generate a series of building Design Alternatives that were essentially individual OSWs, each processed by the OpenStudio CLI.

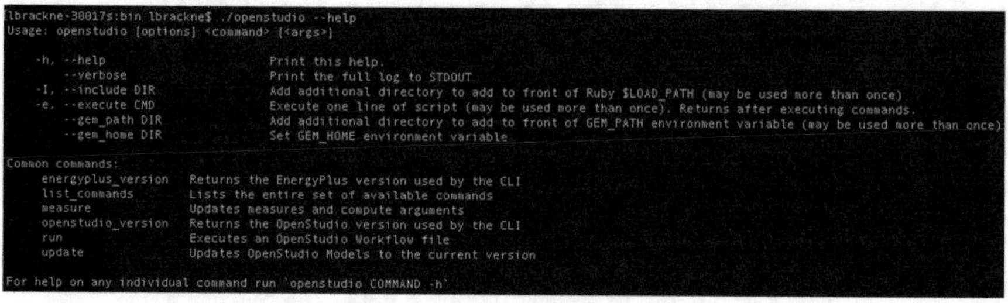

```
lbrackne-30017s:bin lbrackne$ ./openstudio --help
Usage: openstudio [options] <command> [<args>]

    -h, --help                  Print this help.
        --verbose               Print the full log to STDOUT
    -I, --include DIR           Add additional directory to add to front of Ruby $LOAD_PATH (may be used more than once)
    -e, --execute CMD           Execute one line of script (may be used more than once). Returns after executing commands.
        --gem_path DIR          Add additional directory to add to front of GEM_PATH environment variable (may be used more than once)
        --gem_home DIR          Set GEM_HOME environment variable

Common commands:
    energyplus_version          Returns the EnergyPlus version used by the CLI
    list_commands               Lists the entire set of available commands
    measure                     Updates measures and compute arguments
    openstudio_version          Returns the OpenStudio version used by the CLI
    run                         Executes an OpenStudio Workflow file
    update                      Updates OpenStudio Models to the current version

For help on any individual command run 'openstudio COMMAND -h'
```

**Fig. 9.36** OpenStudio CLI help text

In Chap. 7, we explored PAT's ability to generate hundreds or thousands of OSWs automatically. While each individual OSW was still processed by the CLI, they were generated by another part of the SDK called the OpenStudio "Meta" CLI – so-called because it is a CLI that generates inputs for another CLI. If the CLI is the cook, then the Meta CLI is the chef, overseeing all of the individual cooks in the kitchen to satisfy a restaurant full of hungry diners.

Whereas the CLI takes instruction from an OSW, the Meta CLI uses an OpenStudio Analysis (OSA) file. An OSA describes the characteristics of an algorithm that will autonomously create individual OSWs. Example analyses include the use of sampling or optimization algorithms that dynamically generate combinations of seeds, Measures, and Measure arguments. Such algorithms enable rapid evaluation of efficiency performance spaces, optimized designs based on multiple objectives, or calibration of energy models against metered consumption data.

OSAs are also JSON files, but specify one or more seed models, one or more weather files, an algorithm and its parameters, and a set of OpenStudio Measures along with their parameters. Unlike workflow Measure arguments that are pre-scribed in an OSW, an OSA can describe ranges, distributions, or sets of arguments that can be explored by the algorithm. The schema for the OSA is currently documented in https://github.com/NREL/OpenStudio-analysis-gem. In Chap. 7, we utilized PAT to generate a number of different OSAs.

The Meta CLI is a compact, cross-platform executable that consumes OSA files and executes them using local, cluster, or cloud computing resources.[8] Like the OpenStudio CLI, the Meta CLI can be used to create (more advanced) applications. Figure 9.37 describes the basic concept of using the Meta CLI to execute an analysis. It also illustrates the relationship between the Meta CLI, OpenStudio Server,[9] and parallel computing "workers" running individual instances of the OpenStudio CLI.

As with the CLI, the Meta CLI is invoked from the command line. Invoking it with a -h or --help argument returns basic guidance on using it to start or stop an OpenStudio Server and run an OSA on it (Fig. 9.38).

---

[8] https://github.com/NREL/OpenStudio-server/blob/develop/bin/openstudio_meta
[9] https://github.com/NREL/OpenStudio-server

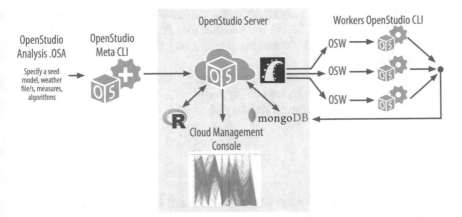

**Fig. 9.37** OpenStudio Meta CLI enabled analysis (*Credit: Marjorie Schott*)

```
lbrackne-30017s:bin lbrackne$ ./openstudio_meta --help
Usage: openstudio_meta [options] <command> [<args>]

    -h, --help                      Print this help.
        --verbose                   Print the full log to STDOUT

Common commands:
        install_gems    Installs the required packaged Gems
        run_analysis    Runs an analysis on an OS Server instance
        start_local     Starts local processes for the OS Server
        start_remote    Starts a remote OS Server
        stop_local      Stops local processes for the OS Server
        stop_remote     Stops a remote OS Server on AWS

For help on any individual command run `openstudio_meta COMMAND -h`
```

**Fig. 9.38**   OpenStudio Meta CLI help text

Use of the OpenStudio Meta CLI outside of PAT is an advanced software engineering topic that can best be explored via documentation located at https://github.com/NREL/OpenStudio-server.

## 9.10   Additional Exercises (for Advanced Users)

1) Continue learning about OpenStudio Measures by creating additional variations of existing Measures or write an entirely new one.
2) Explore the OpenStudio CLI by creating some simple calculators that take input from the user, write one or more OSW files, and call the CLI. A few simple ideas include:

- A before and after retrofit calculator that involves two OSW files – both with the same seed Model and weather file. The two workflow files would only differ in the EE Measures and/or Measure arguments applied in the retrofit case.
- A simple calculator that assesses the energy savings potential of a new EE technology across many building types by utilizing an empty seed model and calling the DOE Prototype Building Measure as the first step in a workflow.

# References

https://bcl.nrel.gov/search/site/GLHEPro?f[0]=bundle%3Anrel_measure
http://codebeautify.org/xmlviewer
https://hvac.okstate.edu/glhepro/overview
https://github.com/NREL/OpenStudio-analysis-gem
https://github.com/NREL/OpenStudio-server
https://github.com/NREL/OpenStudio-server/blob/develop/bin/openstudio_meta
https://nrel.github.io/OpenStudio-user-documentation/reference/command_line_interface
https://nrel.github.io/OpenStudio-user-documentation/reference/command_line_interface/#osw-structure
http://nrel.github.io/OpenStudio-user-documentation/reference/measure_writing_guide
https://www.ruby-lang.org/en/documentation/quickstart/

# Correction to: Building Energy Modeling with OpenStudio

Larry Brackney, Andrew Parker, Daniel Macumber, and Kyle Benne

**Correction to:**
**L. Brackney et al., *Building Energy Modeling with OpenStudio*,**
**https://doi.org/10.1007/978-3-319-77809-9**

This book was inadvertently published without the online supplementary files for chapters 2 through 9. This has now been updated accordingly.

---

The updated online version of this book can be found at
https://doi.org/10.1007/978-3-319-77809-9

The updated online versions of these chapters can be found at
https://doi.org/10.1007/978-3-319-77809-9_2
https://doi.org/10.1007/978-3-319-77809-9_3
https://doi.org/10.1007/978-3-319-77809-9_4
https://doi.org/10.1007/978-3-319-77809-9_5
https://doi.org/10.1007/978-3-319-77809-9_6
https://doi.org/10.1007/978-3-319-77809-9_7
https://doi.org/10.1007/978-3-319-77809-9_8
https://doi.org/10.1007/978-3-319-77809-9_9

# Resources

This book is intended as a structured guide to beginning energy modeling using OpenStudio and EnergyPlus. The scope of building energy modeling is vast, and it is not possible to cover all of the features of these continually evolving tools, let alone the nuances of systems modeling that can take years to master. Fortunately, many online resources can assist you as you continue to practice what you have begun to learn in this book.

The central hub for information on OpenStudio is the project home page at https://openstudio.net. In addition to software downloads, you will find user documentation, tutorials, YouTube videos, a developer section, and documentation to the SDK - crucial for those who wish to develop their own Measures and OpenStudio applications. Figure A.1 is the main OpenStudio user documentation page featuring links to some of the online materials available at the site.

The site also contains a link to https://unmethours.com, an online forum supported by a vibrant community of energy modeling practitioners, researchers, and tool developers (Fig. A.2). Unmet Hours uses a searchable question and answer format, and is a tremendous resource for both novice and expert energy modelers.

> **Tip**: If you get stuck on a modeling problem, or think you've stumbled across a software bug, Unmet Hours is the place to go, **but**...
> ... please remember to search for your problem before posting. You may often find your question has already been asked and answered.

As we have noted many times throughout this book, OpenStudio hews very closely to EnergyPlus' component hierarchy. As such, OpenStudio users can and should take advantage of EnergyPlus' extensive documentation. We have included footnoted URLs to both the EnergyPlus Input Output Reference Guide as well as the EnergyPlus Engineering Reference throughout the text for good reason. Both references may be accessed at https://energyplus.net as shown in Fig. A.3.

© Springer International Publishing AG, part of Springer Nature 2018
L. Brackney et al., *Building Energy Modeling with OpenStudio*,
https://doi.org/10.1007/978-3-319-77809-9

**Fig. A.1**  OpenStudio documentation web page

We leave you with a few troubleshooting tips for common problems that new modelers frequently stumble into.

## Common Simulation Failures

Even the most experienced energy modelers can be faced with the dreaded simulation failure. Outright failures to simulate are generally well documented in the *eplusout.err* file that EnergyPlus creates. Don't be afraid to open this file and scroll through the contents. It won't bite, and the answers you seek are often there in plain English.

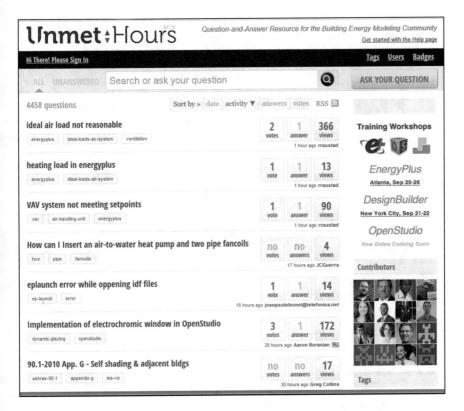

**Fig. A.2**  The Unmet Hours web page

Common errors that may result in a simulation failure include:

**Weather Related**
- **Q**: Did you include an EPW weather file?
- **A**: Add one.
- **Q**: Did you include a DDY design day file?
- **A**: If you expected systems to be auto-sized then add one.

**Envelope Related**
- **Q**: Were Constructions assigned to all surfaces in your model?
- **A**: Use the Geometry Previewer to render by construction and check.
- **Q**: Are interior surfaces matched with other interior surfaces?
- **A**: Use the Geometry Previewer to render by boundary condition and check.

**Thermal Zone Related**
- **Q**: Does your model include at least one Thermal Zone and Air Terminal?
- **A**: EnergyPlus requires at least one of each to simulate. Add them.
- **Q**: Do your Thermal Zones have associated heating and cooling setpoint schedules?
- **A**: EnergyPlus requires heating and cooling schedules to drive HVAC systems. Add them.

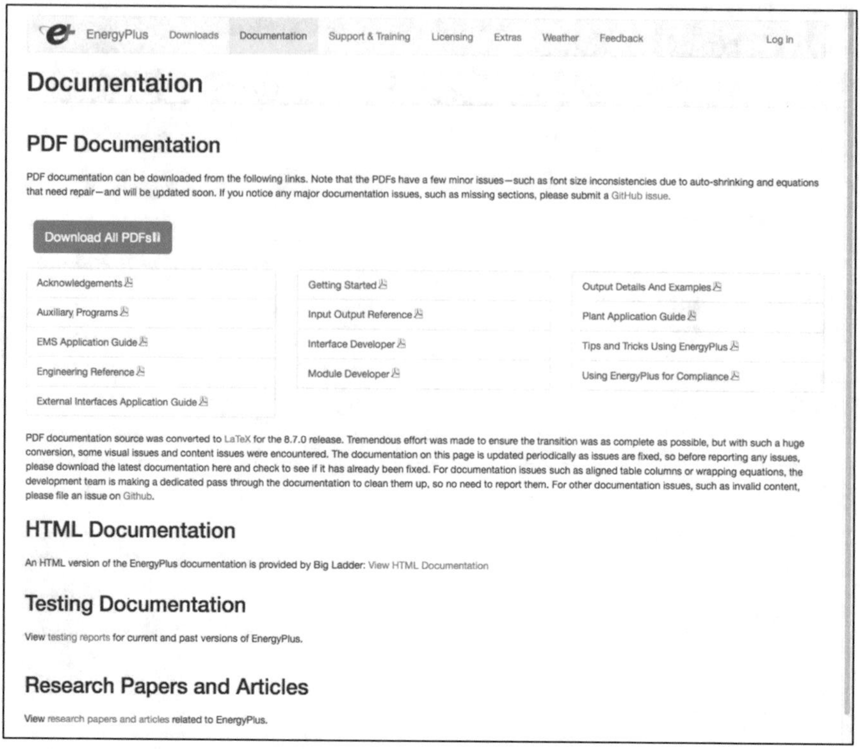

**Fig. A.3** EnergyPlus documentation web page

# Simulation "Soft" Failures

Just because a simulation ran doesn't mean it necessarily ran "correctly." Throughout the text, we have attempted to impress upon the reader the importance of reviewing simulation results and questioning their validity through application of experience, common sense, and intuition. Energy modeling, like any type of modeling, is a garbage in/garbage out process. This means that no matter how good the engine is, incorrect input will lead to incorrect output.

- Always double-check your inputs for correctness.
- Check detailed simulation results to make sure that what you expect to happen in your model is actually happening. This includes annual, monthly, and time series results.
- When appropriate, compare your energy results to known benchmarks such as the DOE Reference Buildings to make sure that your results are in the right ballpark.
- Try incorporating some of the Automated QAQC (Quality Assurance Quality Checking) Reporting Measures from the BCL in your workflow to quickly "sanity test" your results.

For simulation results that don't seem quite right, consider the following:

**End Use Related**
- **Q**: Are your Spaces assigned to a Thermal Zone?
- **A**: If they aren't, then their Loads will not be reflected in the building's performance. Assign them.
- **Q**: Did you assign Loads to your Spaces?
- **A**: Check the Spaces (■) Tab for inherited or manually assigned loads. If any are missing, add them.

**Excessive Unmet Hours**
- **Q**: Were your Thermal Zone design sizing parameters consistent with Air Loop HVAC targets?
- **A**: Review Sect. 5.2.2. Note that Zone sizingparameter and Loop target mismatch can result in undersized systems.
- **Q**: If you attached multiple Thermal Zones to a Loop, were their sizing parameters set consistently?
- **A**: Again, review Sect. 5.2.2. Ensure that all sizing targets are consistent.

## PAT Issues

PAT's ability to run locally or in the cloud is unique, but can pose some challenges for new users. The following tips can help avoid or correct common problems:

**Local Server Won't Start**
- **Q**: Did you wait at least a minute before checking to see if the server is available?
- **A**: PAT can take up to a minute to start processes required for simulation. Check back in a bit.
- **Q**: Is the server not available after a minute or more?
- **A**: The following steps can help release a local server instance that has gotten stuck:

Make sure all the server processes are running. Open your task or process manager and look for ruby processes. If the server is running, you should see several – roughly equal to the number of cores on your machine + one (Fig. A.4).

Also, check for a process named "mongod." This is the database that stores simulation results.

Check to see if the server is responsive in your browser by going to http://localhost:8080. If you get an error saying the site can't be reached, things aren't running correctly, and the following action should be taken:

- Under PATs Windows Menu, select "Server Troubleshooting Tools." This will bring up the window shown in Fig. A.5.

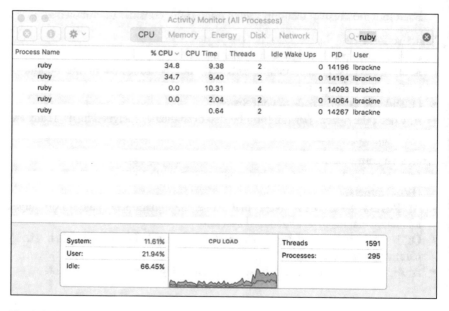

Fig. A.4  PAT server running on a four core laptop

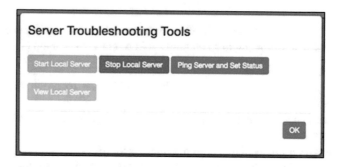

Fig. A.5  PAT server troubleshooting tools

- Click the [Stop Local Server] Button. This should cause the ruby and mongod processes to disappear from your task manager or activity monitor. If they persist, use the manager/monitor to manually quit them.
- Quit PAT
- Go to your Project directory and manually delete the files highlighted in Fig. A.6.
- Restart PAT. In nearly all cases, simply waiting for the server to start properly before running an analysis will take care of things.

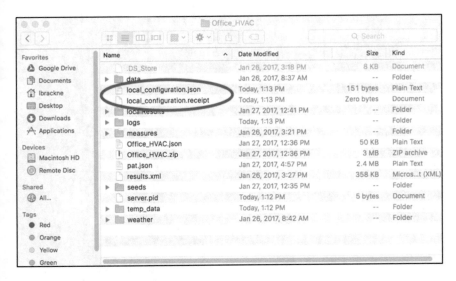

**Fig. A.6** PAT server troubleshooting tools

## Local Server Starts But Analysis Won't Run

- **Q**: Why does PAT seem "stuck" running when I try to run my project?
- **A**: PAT is unable to run Projects from network drives. Make sure your Project files are stored locally.

## Measure Manager in PAT is Empty

- **Q**: I open the Measure Manager in PAT to add Measures to my Project but it is empty. What's wrong?
- **A**: In some rare instances, OpenStudio's Measure Manager process can be orphaned.

  - Quit PAT
  - Use your task or process manager to look for a stray process named "openstudio."
  - Force them to quit.
  - Restart PAT.
  - You should see Measure content again.

## AWS Server Won't Start

- **Q**: I am trying to run PAT on the cloud via AWS, but the server isn't starting.
- **A**: Did you enter your AWS credentials correctly?
- **A**: Check the EC-2 console for system outage alerts.
- **A**: Monitor your EC-2 instances as shown in Chap. 7 to see if your cluster is starting.
- **A**: Did you use the correct AMI for your version of OpenStudio?

# Index

© Springer International Publishing AG, part of Springer Nature 2018
L. Brackney et al., *Building Energy Modeling with OpenStudio*,
https://doi.org/10.1007/978-3-319-77809-9

Printed in the United States
By Bookmasters